Table of Contents

1 Chapter 1: Introduction to Egg Laying

Egg laying is a process seemingly simple, yet it holds within it a universe of biological complexity, making it an ideal subject for meticulous study. To begin, it's crucial to grasp the multifaceted nature of this phenomenon. We're talking about more than just chickens laying eggs; it's a fundamental aspect of reproduction across a vast spectrum of species, from insects to amphibians, and even certain reptiles. Each species has evolved unique strategies and biological mechanisms that govern when, where, and how they lay their eggs. Some, for example, meticulously construct nests, providing a safe haven for their offspring, while others deposit their eggs in less protected environments, relying on sheer numbers or camouflage for survival. Consider, for instance, the intricate egg-laying behaviors of some insects. Certain species of butterflies lay their eggs exclusively on specific host plants, ensuring that their larvae have immediate access to the food they need to survive. Others, like certain parasitic wasps, inject their eggs directly into the bodies of other insects, turning them into living incubators for their young. These fascinating adaptations highlight the diversity and complexity of egg laying in the natural world.

Now, let's dive into the various factors influencing egg deposition. It's not merely a matter of biological readiness; a symphony of environmental cues and internal physiological processes orchestrates this crucial event. Age plays a significant role, as younger organisms may not yet be fully reproductively mature, while older ones may experience a decline in fertility. Diet is another critical factor. A well-nourished organism is far more likely to produce healthy, viable eggs than one that is malnourished. The availability of specific nutrients, such as calcium, is often essential for eggshell formation. Furthermore, environmental conditions such as temperature, humidity, and light can profoundly impact egg laying. Many species are highly sensitive to these factors, and deviations from their preferred conditions can disrupt their reproductive cycles. For example, some insects only lay eggs during specific seasons when temperatures are optimal for larval development. Imagine trying to study the egg-laying behavior of a particular insect species in a lab setting where the temperature is consistently too cold. You might observe a complete cessation of egg laying, leading to inaccurate conclusions about the species' reproductive capabilities.

The benefits of studying egg laying in a controlled environment cannot be overstated. Natural habitats are often rife with uncontrollable variables, making it difficult to isolate the specific factors that influence egg deposition. In the wild, predators, competitors, and fluctuating environmental conditions can all impact egg laying rates and survival. By contrast, a controlled environment offers the opportunity to eliminate these confounding variables and create conditions where every aspect of the organism's surroundings is carefully regulated. This allows researchers to manipulate specific factors, such as temperature or diet, and observe their direct impact on egg laying. Such precision is simply unattainable in the field, and this control extends to detailed observation. Consider the case of studying the effect of a specific pesticide on eggshell thickness. In the wild, it might be difficult to determine whether a decline in eggshell thickness is due to pesticide exposure or other environmental factors, such as poor nutrition. However, in a controlled environment, you can expose organisms to varying concentrations of the pesticide and directly measure its effect on eggshell thickness, providing definitive evidence of its impact.

Moreover, studying egg laying in a controlled environment facilitates detailed observation. Under a microscope, we can observe the formation of the egg, the development of the embryo, and even the hatching process. Such insights into the intricate biological mechanisms that govern egg laying are impossible to obtain in the field. Imagine trying to observe the microscopic changes occurring within a developing egg in the wild. You would be constantly battling against environmental distractions, such as wind, rain, and predators. In a controlled environment, you can observe the egg under ideal conditions, using sophisticated imaging techniques to capture every detail of its development. This level of observation is crucial for understanding the underlying biological processes that drive egg laying. Fur-

thermore, such experiments hold a significant weight when it comes to ethology. Observing animals behaviors, and their changes, while they have interaction with the eggs helps in better understanding the animal behaviour.
In conclusion, understanding egg laying requires a multifaceted approach, combining knowledge of species-specific behaviors with an appreciation for the environmental and physiological factors that influence this critical process. And the ability to precisely control these factors in a lab setting not only improves the accuracy of research, but it makes the experiments themselves easier to perform, and the data much more readily available.

1.1 Egg Laying: An Overview

Egg laying, at its core, is the reproductive process that underpins the continuation of numerous species, particularly within the insect world. It's more than a simple act of depositing eggs; it's a complex interplay of physiological, environmental, and behavioral factors meticulously orchestrated to ensure the survival of the next generation. For anyone venturing into the study of egg laying, understanding this fundamental process is crucial before delving into the intricacies of controlled experimentation. The oviposition behaviour itself, the selection of the laying site, the number of eggs deposited, the protection provided to the eggs, all these aspects are finely tuned by evolution to maximize reproductive success in the specific environment where the organism lives.

Now, let's break down the key components involved in egg laying. Firstly, the physiological aspect involves the development and maturation of oocytes within the female reproductive system. This process is intricately linked to hormonal regulation, nutritional intake, and overall health. Think of it as a carefully calibrated engine: if one part is malfunctioning, the entire system is compromised. In many insects, for example, specific proteins derived from their diet are essential for yolk formation, providing the developing embryo with the necessary nutrients for its initial growth. Secondly, environmental cues play a crucial role in triggering egg laying. Temperature, humidity, light cycles, and the availability of suitable substrates all act as stimuli, influencing the timing and location of egg deposition. For instance, some insects only lay eggs when temperatures reach a certain threshold, ensuring that the offspring hatch under optimal conditions. Others are highly selective in their choice of oviposition sites, seeking out specific plants or substrates that provide food and shelter for the developing larvae. You will see that this interplay between the organism and its surrounding can deeply influence its behaviour, and eventually its evolution and adaptation.

Then comes the behavioural aspect, which is perhaps the most fascinating. The act of egg laying is often preceded by elaborate courtship rituals, mate selection, and nest building. Females may assess the quality of potential mates based on various signals, such as pheromones or visual displays, ensuring that their offspring inherit advantageous traits. Furthermore, the selection of oviposition sites is not random; females of-

ten exhibit a remarkable ability to assess the suitability of a particular location for their offspring's survival. They might consider factors such as the presence of predators, the availability of food resources, and the microclimate of the area. This intricate decision-making process highlights the sophisticated cognitive abilities of many insects. If you observe insects and invertebrates in their natural habitat you can notice how carefully they choose the place to leave their eggs and offspring.
Studying egg laying in a controlled setting offers a unique opportunity to dissect these complex interactions and gain a deeper understanding of the underlying mechanisms. By manipulating environmental parameters, such as temperature, humidity, and light, researchers can investigate their effects on egg production, hatch rate, and larval development. Furthermore, controlled experiments allow for the isolation and study of specific genetic factors that influence egg laying behavior. For instance, researchers can use genetic techniques to knock out or overexpress certain genes, and then observe the resulting changes in egg laying behavior. I think that the most useful method is to combine experimentation with a good understanding of the natural life and ecology of the studied insect.
Furthermore, the controlled environment provides an ideal setting for studying the effects of various environmental toxins and pollutants on egg laying. By exposing insects to different concentrations of these substances, researchers can assess their impact on egg production, hatchability, and larval survival. This type of research is crucial for understanding the ecological consequences of environmental contamination and for

developing strategies to mitigate its harmful effects. I have personally noticed that the higher is the controlled environment, the better are the results, and the less contamination you find when you extract the data.

Finally, understanding the nuances of egg laying is not just an academic exercise; it has important practical applications as well. For example, in agriculture, knowledge of egg laying behavior can be used to develop more effective pest control strategies. By disrupting the egg laying process, it may be possible to reduce pest populations without resorting to harmful pesticides. Similarly, in conservation biology, understanding the factors that influence egg laying can help us to protect endangered species by ensuring that they have access to suitable breeding habitats. This also involves understanding the ecological networks the insect is part of. You can't just pretend you understand egg laying if you don't understand also the other steps of the insect life and relationships.

I would say, though, that there are also challenges and limitations when studying egg laying in controlled environments. The artificial conditions may not fully replicate the complexities of the natural world, and the results obtained in the lab may not always translate directly to field conditions. Therefore, it is important to complement laboratory studies with field observations and experiments to gain a more comprehensive understanding of egg laying behavior. The more you compare your data with the natural world, the better they are.

Also, keep in mind that the manipulation of insect egg laying can raise ethical concerns, particularly when dealing with endangered or threatened species. It is important to adhere to strict ethical guidelines and to minimize any potential harm to the insects being studied. Obtain necessary permits and approvals from relevant regulatory agencies before conducting any experiments involving egg laying. Prioritize animal welfare and use humane methods for handling and housing insects. So you will ensure a very ethical experiment.

In conclusion, egg laying is a multifaceted process that plays a critical role in the life cycle of many insects. By studying egg laying in controlled environments, researchers can gain valuable insights into the underlying mechanisms that govern this process, leading to a better understanding of insect biology and ecology. It is a fascinating field of study with important implications for agriculture, conservation, and human health. As you embark on your journey of exploring egg laying, remember to approach your work with curiosity, rigor, and a deep respect for the natural world.

1.2 Factors Influencing Egg Deposition

The act of egg deposition, while seemingly simple, is a culmination of various internal and external factors intricately woven together. These factors can significantly influence the frequency, timing, and even the quality of eggs laid. Understanding these influences is vital for anyone seeking to study or manage egg-laying species, particularly in controlled environments. To get started, let's dig a little deeper.

One of the most fundamental influences is nutrition. A

female's nutritional state directly affects her ability to produce eggs. Adequate protein is crucial, as it forms the building blocks of the egg itself. Insufficient protein intake can lead to a reduced number of eggs, smaller egg size, or even a complete cessation of egg laying. Similarly, the availability of essential minerals like calcium and vitamins, such as vitamin D, are non-negotiable for eggshell formation and overall reproductive health. A deficiency in these nutrients can result in thin-shelled eggs, which are prone to breakage and desiccation. Furthermore, lipids play a key role in yolk development, so a balanced diet including sufficient fats is necessary for producing viable offspring. It's not just the presence of these nutrients but also their ratios. An imbalance can interfere with the absorption and utilization of other essential components. Think of it like baking a cake – you can't just throw in flour and expect it to rise perfectly; you need the right amount of sugar, butter, and eggs. Similarly, egg production depends on the right balance of nutrients.

Then comes the critical influence of environmental conditions, and temperature is a prime factor. Most egg-laying species are ectothermic, meaning their body temperature is largely dependent on the external environment. Optimal egg production generally occurs within a specific temperature range, varying across species. Too cold, and the metabolic processes required for egg formation slow down, potentially halting egg laying altogether. Too hot, and the female may experience heat stress, negatively impacting her reproductive system and reducing egg output. Humidity also plays a significant role. Insufficient humidity can lead to egg desiccation, rendering them inviable, while excessive humidity can create conditions favorable for fungal growth, increasing the risk of infection and egg spoilage. Light is another often-overlooked factor. Photoperiod, the duration of light exposure each day, can act as a powerful cue for initiating and regulating reproductive cycles. Many species are seasonal breeders, with egg laying occurring primarily during periods of longer day length. This is because light influences the release of hormones that stimulate reproductive development and behavior. Consider mimicking natural light cycles as much as possible to encourage a healthy laying cycle.

Now, let's discuss the species-specific factors, starting with age. Egg-laying ability typically varies with age. Young females may lay fewer eggs or produce smaller eggs compared to mature females. Conversely, older females may experience a decline in egg production as their reproductive systems begin to age. Understanding the reproductive lifespan of your chosen species is important for interpreting egg-laying data and optimizing breeding strategies. Genetics also play a significant role. Some species have naturally higher egg-laying rates than others, and there can even be variation within a species due to genetic differences. Selective breeding programs can be used to enhance egg production in certain species, but it's essential to consider the ethical implications of such practices. Mating history can influence egg deposition as well. In species that store sperm, a single mating may be sufficient to fertilize multiple clutches of eggs. However, in other species, repeated mating may be necessary to maintain egg fertility. Understanding the mating system of your species is essential for managing reproduction.

In addition to these elements, stress, whether physical or psychological, can significantly impact egg deposition. Crowded conditions, aggressive interactions with other individuals, or exposure to predators can all induce stress, leading to reduced egg production or even egg retention. I found that providing adequate space, minimizing disturbances, and creating a secure environment can help to reduce stress and promote healthy egg laying. For example, if you are studying insects, be mindful of the substrate they prefer to lay their eggs on. Some species are very specific about their laying sites, and providing the correct substrate is essential for inducing egg deposition. Others may be more sensitive to vibrations or noise.

Lastly, there are the indirect influencers, disease and parasites. Infections can compromise a female's health and reduce her ability to produce eggs. Parasites can also drain resources and interfere with reproductive processes. Regular health checks and preventative measures can help to minimize the risk of disease and parasite infestations. The presence of environmental contaminants can affect the eggs themselves. Exposure to pesticides, heavy metals, or other pollutants can disrupt hormone function, leading to reduced egg production, eggshell abnormalities, or developmental problems in the offspring. Be careful to filter the water.

In summary, egg deposition is a complex process influenced by a multitude of factors, from nutrition and environmental conditions to species-specific traits and stress levels. By understanding these influences and carefully controlling them in a research environment, we can gain valuable insights into the underlying mechanisms that regulate egg laying and improve our ability to manage egg-laying species effectively. Remember, meticulous observation and careful data collection are essential for unraveling the mysteries of egg deposition. These processes are the key components to your success in researching egg-laying.

1.3 Controlled Environment Benefits

Consider, for a moment, the intricate dance of variables that influence egg deposition in any given species. Out in the wild, these variables – temperature fluctuations, unpredictable food availability, fluctuating humidity, the presence of predators – create a cacophony of factors that can mask the true underlying mechanisms driving reproductive behavior. Herein lies the profound advantage of a controlled environment. By meticulously regulating these external elements, we can isolate and scrutinize the specific triggers and physiological pathways that govern egg laying, allowing us to finally understand how these organisms work in a precise and reproducible way.

Think of a thermostat. It maintains a constant temperature in your home, regardless of the weather outside. A controlled environment for studying egg laying works similarly, only on a far more complex and delicate scale. This precision is paramount when trying to understand the subtle effects of different stimuli on egg production. We can precisely control lighting schedules, mimicking dawn and dusk with unparalleled accuracy, and thereby understand the photoperiod's influence. Similarly, the composition and quantity of food available can be meticulously controlled and altered, enabling the tracking of nutritional influences. Imagine the possibilities: precisely measuring the impact of specific nutrients on egg viability or the timing of oviposition.

What are the other advantages of these systems? Consistent data collection becomes vastly simpler. When external factors are held constant, observed variations in egg laying can be directly attributed to the manipulated variables. This leads to significantly more robust and reliable data, minimizing the impact of random noise and maximizing the statistical power of your analysis. For instance, you can confidently assess the effect of a specific hormonal manipulation on egg laying when you know that the ambient temperature, humidity, and light levels are not confounding your results. Such clarity simply is not possible in a natural setting.

Let's not forget the ethical considerations. Controlled environments, when designed and operated responsibly, can actually enhance animal welfare. They offer protection from predators, harsh weather conditions, and resource scarcity. By providing consistent access to food, water, and appropriate habitat, we minimize stress and maximize the overall health of our study subjects. This not only benefits the animals themselves but also improves the quality and reliability of our scientific findings. A happy, healthy animal is more likely to exhibit normal reproductive behavior, leading to more accurate and representative results.

However, it's important to recognize that a controlled

environment isn't a magic bullet. It's a tool, and like any tool, it must be used carefully and thoughtfully. One potential pitfall is oversimplification. While controlling external factors is beneficial, it's crucial to ensure that the environment remains stimulating and enriching enough to promote natural behaviors. A sterile, featureless enclosure can be just as detrimental as an unpredictable natural habitat. Therefore, it's imperative to provide appropriate substrates for egg laying, opportunities for social interaction (where appropriate), and environmental enrichment that mimics aspects of the species' natural environment.

Moreover, a controlled environment can sometimes create artificial selection pressures. For example, if a certain genotype happens to thrive particularly well under the controlled conditions, it may become overrepresented in the population over time. This can skew results and limit the generalizability of the findings to natural populations. To mitigate this risk, it's essential to maintain genetic diversity within the study population and to periodically introduce individuals from outside sources to prevent inbreeding and genetic drift.

Consider also the logistics. Establishing and maintaining a controlled environment can be resource-intensive. It requires specialized equipment, dedicated space, and ongoing monitoring. Temperature and humidity control systems, lighting setups, and air filtration systems all require significant investment and maintenance. Furthermore, meticulous record-keeping is essential to ensure the integrity of the experimental conditions. Daily logs of temperature, humidity, light levels, and other relevant parameters are crucial for identifying and ad-

dressing any deviations from the target conditions.
Now, how can you apply these benefits in a practical setting? Start by clearly defining your research question and identifying the specific variables that you want to manipulate and control. This will help you determine the necessary equipment and environmental parameters. Next, carefully research the species you intend to study and design your controlled environment to meet its specific needs. Consider factors such as temperature preferences, humidity requirements, substrate preferences, and social behavior.
Ultimately, the benefits of a controlled environment extend far beyond simply simplifying data collection. It allows us to dissect the complex interplay of factors that govern egg laying, providing unprecedented insights into the physiological, behavioral, and evolutionary mechanisms at play. By carefully designing and managing these environments, we can unlock a deeper understanding of these fascinating processes and pave the way for new discoveries in zoology, developmental biology, and related fields. It is a commitment to precision that unlocks the secrets of life, one meticulously controlled experiment at a time, revealing the wonders of nature in a way that would be impossible in the uncontrolled chaos of the outside world. It's a powerful method to uncover the mysteries that natural environments often obscure.

2 Chapter 2: Setting Up the Controlled Environment

Setting up the controlled environment is paramount to achieving reliable and reproducible results in egg-laying and larval transformation studies. It's not simply about confining organisms; it's about meticulously crafting an ecosystem tailored to their specific needs. This requires a thoughtful approach to enclosure selection, environmental control, and monitoring, all with the aim of minimizing extraneous variables and maximizing data quality. Consider this stage as building the foundation for everything that follows, a meticulously planned habitat that allows for focused study.

Choosing the right enclosure begins with understanding the natural history of your chosen species. Are they terrestrial, aquatic, or arboreal? What is their typical range of movement? The enclosure must provide adequate space for the organisms to move, feed, and reproduce without experiencing undue stress. A cramped or unsuitable environment can inhibit egg laying, distort larval development, and ultimately compromise the validity of your experiment. Material is another crucial aspect. Glass aquariums are excellent for aquatic species, offering clear visibility and ease of cleaning. Plastic containers are lighter and more durable, suitable for terrestrial insects. However, it's essential to ensure that the plastic is non-toxic and doesn't leach harmful chemicals into the environment. Ventilation is essential. Stagnant air can lead to the build-up of harmful gases and pathogens. Choose enclosures with built-in ventilation or modify existing ones to provide adequate airflow. However, be mindful of the size of the openings, preventing escape or the entry of unwanted pests. Consider the lighting as well. Natural light can be beneficial, but direct sunlight can cause overheating. Artificial lighting can be used to mimic natural day-night cycles. The type of light (full-spectrum, LED, etc.) should be appropriate for the species and the experimental goals. Remember that the goal is to create a stable and predictable environment.

Temperature is a critical factor influencing egg laying and larval development. Many invertebrates are ectothermic, meaning their body temperature is largely dependent on the surrounding environment. Fluctuations in temperature can disrupt metabolic processes, delay development, or even lead to mortality. Maintaining a stable temperature within the optimal range for your species is crucial. Small incubators or climate-controlled chambers are ideal for precise temperature control. These units typically have built-in heating and cooling systems, allowing you to set and maintain a constant temperature. Thermostats and temperature sensors are also useful for monitoring temperature within the enclosure. These devices provide real-time readings, allowing you to make adjustments as needed. When controlling temperature, consider the potential for thermal gradients within the enclosure. Place temperature sensors in different locations to ensure that the temperature is uniform throughout. Pay attention to seasonal variations as well. Even in a controlled environment, external temperature changes can affect the internal temperature of the enclosure. Adjust your settings accordingly to compensate for these fluctuations. Accurate temperature control is not just about maintaining a specific degree; it's about creating a consistently stable environment.

Humidity is just as important as temperature. Many invertebrates require a specific level of humidity for proper egg laying and larval development. Too little humidity can cause eggs to dry out, while too much can promote the growth of mold and bacteria. A hygrometer is essential for monitoring humidity levels. These devices measure the amount of moisture in the air, providing you with the information you need to make adjustments. There are several ways to increase humidity in an enclosure. Misting the enclosure with water is a simple and effective method. However, avoid over-misting, as this can lead to excessive moisture and the growth of pathogens. A humidifier can be used to maintain a constant level of humidity. These devices release water vapor into the air, creating a humid environment. A substrate like vermiculite or peat moss can help retain moisture and increase humidity. Keep the substrate moist, but not waterlogged. Conversely, decreasing humidity can be achieved through ventilation. Increase airflow to reduce moisture levels. Desiccants, such as silica gel, can also be used to absorb excess moisture. Aim to maintain a stable humidity level within the optimal range for your species. Monitor humidity levels regularly and make adjustments as needed. Remember that the ideal humidity level will vary depending on the species. Research the specific requirements of your chosen organism to ensure that you're providing the right conditions. A balance needs to be struck, and carefully monitor how your selected population responds to the designed habitat, and be ready to make any needed adjustment.

Beyond the basics of temperature and humidity, consider other environmental factors that may influence egg laying and larval development. Light is an essential cue for many invertebrates. Provide a natural day-night cycle using artificial lighting or by placing the enclosure near a window. Avoid exposing the organisms to direct sunlight, which can cause overheating. Substrate is also important. Provide a suitable substrate for egg laying and larval development. The type of substrate will vary depending on the species. Soil, sand, leaf litter, or even paper towels can be used. Food is critical. Provide a consistent and readily available food source for the larvae. The type of food will depend on the species. Algae, plant matter, or commercially available larval food can be used. Water is of course, a necessity. Provide a source of clean water for the organisms to drink. A shallow dish of water or a moist sponge can be used. Remember that the goal is to create a stable and predictable environment. Minimizing extraneous

variables will improve the reliability and reproducibility of your results.
Finally, hygiene. Maintaining a clean and hygienic environment is essential for preventing disease and promoting the health of your organisms. Clean the enclosure regularly to remove waste and debris. Disinfect the enclosure periodically to kill harmful pathogens. Change the substrate regularly to prevent the build-up of bacteria and fungi. Monitor the organisms for signs of disease. Remove any sick or dead individuals immediately to prevent the spread of infection. Wear gloves and wash your hands thoroughly after handling the organisms or the enclosure. Remember that a healthy environment is essential for successful egg laying and larval development studies. Taking the time to set up and maintain a controlled environment is an investment that will pay off in the long run. By providing the right conditions, you can increase the reliability of your data, minimize stress on your organisms, and ultimately gain a deeper understanding of their biology. This chapter is about more than just setting up a box; it's about cultivating a thriving, controlled world for your research.

2.1 Choosing the Right Enclosure

The first critical decision in embarking on egg-laying and larval transformation studies hinges on the proper selection of an enclosure. More than a mere container, the enclosure is the microcosm within which your entire experiment will unfold. Think of it as a carefully designed stage upon which the drama of life cycles will play out, demanding considerable attention to detail.

The size of your enclosure should correlate directly with the species you intend to study and the duration of your experiment. Cramped quarters can lead to stress, inhibited reproduction, and skewed results, whereas an excessively large enclosure can make observation and data collection needlessly difficult. Consider the natural habitat and behavior of your selected species. Are they territorial? Do they require vertical climbing space? The enclosure should facilitate, not hinder, their natural behaviors. For small insects like Drosophila melanogaster, a simple, clear plastic vial might suffice, while larger insects or amphibians may require terrariums or custom-built enclosures with ample room for movement and environmental gradients. I have personally seen studies fail because researchers underestimated the activity levels of their subjects, leading to injuries and disrupted egg-laying patterns.

Material selection is another cornerstone of enclosure design. Glass, plastic, and metal are common choices, each possessing its own set of advantages and disadvantages. Glass offers excellent visibility and is easily sterilized, but it is also heavy and prone to breakage. Plastics are lightweight and durable but can be susceptible to scratching and may leach chemicals, potentially impacting your experimental organisms. Metal cages, often used for larger insects, must be non-toxic and resistant to corrosion. Consider the specific needs of your species and the long-term implications of your material choice. I once used the wrong type of sealant on a plastic enclosure, resulting in the death of an entire colony due to the release of toxic fumes. Learn from my mistakes: always research the compatibility of materi-

als with your chosen species.
Ventilation is absolutely vital to prevent the buildup of harmful gases and maintain optimal humidity levels. Insufficient ventilation can lead to stagnant air, promoting fungal growth and the accumulation of ammonia from waste products, which can be detrimental to egg viability and larval survival. Conversely, excessive ventilation can dry out the enclosure, creating an inhospitable environment. Strike a balance by incorporating appropriate vents or using breathable fabrics for enclosure covers. The size and placement of vents should be carefully considered to ensure proper airflow without creating drafts. I recommend monitoring humidity levels regularly using a hygrometer and adjusting ventilation accordingly.
Substrate selection should mimic the natural environment of your chosen species as closely as possible. Different species have different substrate requirements. Some may require soil for burying eggs, while others prefer leaf litter, moist paper towels, or even bare surfaces. The substrate should be non-toxic, free from pesticides, and capable of retaining moisture without becoming waterlogged. Regularly inspect the substrate for mold or other signs of contamination, and replace it as needed. I've found that using sterilized coconut coir or peat moss provides a good balance of moisture retention and aeration for many species.
Temperature regulation is an essential aspect of enclosure design. Use a reliable thermostat and heating element to maintain the optimal temperature range for your species. Consider the specific temperature requirements for egg incubation and larval development,

as these may differ from the adult stage. Avoid direct contact between the heating element and the substrate to prevent overheating. I prefer using heat mats placed on the side or bottom of the enclosure, rather than heat lamps, to create a more natural temperature gradient.

Humidity control is equally important. Use a humidifier or misting system to maintain the desired humidity level. Consider using a substrate that retains moisture well, such as sphagnum moss or vermiculite. Regularly monitor humidity levels using a hygrometer, and adjust the humidifier or misting system accordingly. Remember that high humidity can promote fungal growth, so proper ventilation is essential. I've discovered that a combination of a shallow water dish and regular misting works well for maintaining humidity for many insect species.

Lighting is another factor to consider. Some species require specific light cycles for proper egg laying and larval development. Use a timer to regulate the light cycle and ensure consistency. Avoid using bright, harsh lights, as these can stress your animals. I recommend using full-spectrum LED lights, which mimic natural sunlight and provide the necessary wavelengths for plant growth (if you are including plants in your enclosure).

Sanitation cannot be overstated. Regularly clean and disinfect your enclosure to prevent the buildup of pathogens. Remove any uneaten food or waste products promptly. Use a mild disinfectant that is safe for your animals, and rinse thoroughly before returning them to the enclosure. I make it a habit to completely disassemble and sterilize my enclosures between experiments to eliminate any lingering contaminants. I know a colleague who lost an entire research project due to a bacterial contamination that could have been easily prevented with proper sanitation protocols.

Finally, consider the aesthetics of your enclosure. While functionality is paramount, a well-designed enclosure can enhance the overall research experience. Use natural materials, such as rocks, wood, and plants, to create a visually appealing and enriching environment for your animals. However, be sure that any materials you use are non-toxic and free from pesticides. I find that adding live plants to my enclosures not only enhances their aesthetic appeal but also helps to regulate humidity and provides a natural food source for some species. Remember that a happy and healthy subject is a more reliable research subject.

In conclusion, the selection of the right enclosure is a multifaceted process that requires careful consideration of numerous factors. By paying attention to size, material, ventilation, substrate, temperature, humidity, lighting, sanitation, and aesthetics, you can create an optimal environment for your egg-laying and larval transformation studies, ultimately leading to more accurate and meaningful results.

2.2 Temperature and Humidity Control

Temperature and humidity are not just background conditions; they are active ingredients in the recipe for successful egg laying and larval transformations. They wield considerable influence over the physiological processes governing these critical life stages. Think of temperature as the engine that drives these processes, while humidity is the oil that keeps everything running smoothly. Neglecting either can lead to significant setbacks in your research.

Maintaining stable temperature is paramount. Enzymes, the biological catalysts that drive biochemical reactions, are incredibly sensitive to temperature fluctuations. Too low, and these reactions grind to a halt, slowing down development and potentially leading to deformities or death. Too high, and these enzymes denature, becoming irreversibly inactive, with equally devastating consequences. The optimal temperature range varies dramatically from species to species, reflecting their evolutionary adaptations to specific climates. For instance, a tropical species might thrive at 28°C, while a temperate species might prefer a cooler 22°C. This means diligent research is essential before you even consider setting up your controlled environment. Don't just guess; consult scientific literature, species-specific care sheets, and the experience of other researchers who have worked with your chosen organism. Once you've identified the ideal temperature range, invest in reliable heating and cooling systems capable of maintaining that range with minimal fluctuation. Thermostatically controlled heating pads, heat lamps, and even climate-controlled incubators can be used, depending on the scale of your setup and the sensitivity of your organisms. A crucial point to remember is that temperature gradients can exist within your enclosure. A heat lamp positioned at one end will create a warmer zone, while the opposite end will be cooler. This can be beneficial, allowing your organisms to thermoregulate by moving to their preferred temperature zone. However, you need to be aware of these gradients and ensure they are not creating extremes that could be detrimental. Regularly monitor temperature at multiple locations within the enclosure using calibrated thermometers. Digital thermometers with remote probes are particularly useful for this purpose, allowing you to track temperature trends over time and identify potential problems before they escalate.

Humidity plays an equally vital role, influencing everything from egg hydration to larval development. Desiccation, or excessive water loss, is a constant threat, especially for delicate eggs and larvae. A dry environment can cause eggs to collapse, preventing proper development, while larvae may struggle to shed their exoskeletons during molting, leading to deformities or death. Conversely, excessively high humidity can create a breeding ground for fungi and bacteria, which can quickly overwhelm your organisms and lead to disease outbreaks. The key, as with temperature, is to strike the right balance. Again, the optimal humidity range is species-specific, so research is crucial. Tropical species generally require higher humidity levels than temperate species, reflecting their adaptation to moist environments. Achieving the desired humidity level can be accomplished through various methods. Simple techniques include placing a shallow dish of water inside the enclosure, allowing it to evaporate gradually. More sophisticated methods involve using humidifiers, which can precisely control the humidity level. Ultrasonic humidifiers are particularly effective, producing a fine mist that is easily absorbed by the air. However, be cautious about over-humidifying the enclosure, as this can lead to condensation on the walls and substrate, creating a favorable environment for pathogens. A hygrometer, an instrument for measuring humidity, is an indispensable tool for monitoring humidity levels within your enclosure. Digital hygrometers are readily available and provide accurate readings. Like thermometers, it's wise to monitor humidity at multiple locations within the enclosure to identify any inconsistencies. Ventilation is another critical factor to consider in humidity control. While maintaining high humidity is important, it's equally important to ensure adequate air circulation to prevent the buildup of stagnant air, which can promote fungal growth. A small fan can be used to gently circulate the air within the enclosure without creating excessive drafts.

Beyond the technical aspects of temperature and humidity control, there's an element of art involved. It's about understanding your organisms' needs and responding to their subtle cues. Are they clustered near the heat source, indicating they're too cold? Are they avoiding the humid areas, suggesting it's too damp? Paying close attention to your organisms' behavior is crucial for fine-tuning your environmental parameters. Maintaining accurate records of temperature and humidity levels, along with observations of your organisms' behavior, is essential for identifying patterns and troubleshooting problems. A sudden drop in egg laying, for example, could be a sign that the temperature or humidity is not optimal. By carefully analyzing your data, you can identify the underlying cause and take corrective action. Don't be afraid to experiment, but always do so in a controlled manner. Make small adjustments to temperature or humidity and observe the effects on your organisms before making further changes. Remember that stability is key. Avoid drastic fluctuations in temperature or humidity, as these can stress your organisms and disrupt their development. A gradual change is always preferable to a sudden shock. The pursuit of understanding egg laying and larval transformations is a journey of discovery, and mastering temperature and humidity control is a vital step along the way. By investing in the right equipment, conducting thorough research, and paying close attention to your organisms' needs, you can create a controlled environment that fosters success and unlocks new insights into the wonders of developmental biology. Think of it not just as a technical challenge, but as an opportunity to create a miniature world where life can flourish under your careful guidance.

3 Chapter 3: Species Selection

The selection of a species is paramount; it sets the foundation for successful and ethical experimentation. But where do you begin? Not every creature that lays eggs is suitable for controlled laboratory studies. We must carefully consider factors like life cycle duration, ease of care, ethical implications, and relevance to the research question. Before you even consider acquiring your specimens, a deep dive into the literature is essential. Understand the natural history of your potential study organism. What are its dietary requirements? What temperature and humidity ranges does it thrive in? How susceptible is it to common diseases? These questions will inform your ability to create a suitable and enriching environment, which will then directly impact the validity of your results. Remember that a stressed animal will not exhibit normal egg-laying behavior, skewing your data.

Now, let's consider life cycle duration. Some species have incredibly long life cycles, spanning months or even years, while others complete their development in a matter of days or weeks. If you're interested in studying the effects of a particular treatment on multiple generations, a species with a shorter life cycle will allow you to collect data more quickly and efficiently. Conversely, if your research question involves intricate developmental processes that unfold over a longer period, a species with a more extended life cycle may be more appropriate. Think about a fruit fly versus a beetle, the difference is abyssal. A species whose life cycle is highly sensitive to environmental changes and can be easily manipulated is desirable. This is because these species are more reactive to environmental changes.

Ease of care is another crucial element to keep in mind. Can you realistically provide the necessary resources and maintain the appropriate environmental conditions for your chosen species? Species with complex dietary requirements or demanding environmental needs can be challenging and expensive to maintain. Species that have already been successfully reared in laboratory settings often have established protocols for feeding, housing, and breeding, making them easier to manage. Furthermore, the species should be tolerant to the lab environment and should not be easily stressed by confinement. If your species is prone to disease in captivity, you may spend more time managing health issues than collecting data, or, worse, get unreliable results due to the sick individuals.

Beyond the practical considerations, ethical considerations loom large. We have a moral obligation to treat our experimental subjects with respect and minimize any potential harm or suffering. The use of vertebrates in research often requires stringent ethical review and justification, while invertebrates may be subject to less regulation. However, this does not mean that we can disregard their welfare. Choose species that are readily available from reputable sources, rather than collecting them from the wild. Wild populations may be vulnerable to over-collection, and removing individuals from their natural habitat can have negative ecological consequences. A suitable choice could be easily bred in captivity or purchased from a certified supplier that adheres to ethical collection practices.

Acquiring and handling your chosen species requires careful planning and attention to detail. Start by contacting reputable suppliers or breeders. Ask them about their rearing practices, health records, and quarantine procedures. Obtain a health certificate to ensure that the animals are free from diseases or parasites. When you receive your specimens, quarantine them in a separate area for a period of time, as recommended by your supplier or a veterinarian specializing in invertebrates, to observe them for any signs of illness. Provide them with optimal environmental conditions and minimize disturbance during this period. Remember, a healthy start is essential for a successful experiment.

When handling your specimens, prioritize their safety and well-being. Use appropriate tools and techniques to minimize stress and avoid injury. For example, use soft-bristled brushes or specialized forceps to handle delicate larvae, and always wear gloves to prevent contamination. Avoid overcrowding the enclosure, and provide adequate space for the animals to move freely. Regularly monitor their behavior for signs of distress, such as erratic movements, reduced feeding, or social withdrawal. If you observe any concerning signs, consult with a veterinarian or experienced invertebrate biologist for guidance. Always prioritize gentle handling and ensure that their natural behaviors are not disrupted excessively.

Moreover, selecting a species that aligns with your research question is crucial. If you are studying the genetic basis of eggshell formation, you will need a species whose genome is well-characterized and amenable to genetic manipulation. If you are interested in the effects of environmental toxins on larval development, you will need a species that is sensitive to those toxins and exhibits clear developmental responses. The more information available about a species, the easier it will be to design and interpret your experiments. Consider the evolutionary relevance of your study organism. Is it a model organism for a particular group of animals? Does it exhibit unique adaptations that are relevant to your research question?

Finally, meticulous record-keeping is essential. Document the source of your specimens, their age, sex, and any relevant medical history. Keep detailed records of their rearing conditions, including temperature, humidity, diet, and lighting. Track their egg-laying behavior, including the number of eggs laid, their size, shape, and hatch rate. And most importantly, keep an ethical perspective in mind when choosing a species for study. By carefully considering all these factors, you can ensure that your research is both scientifically rigorous and ethically responsible. Treat your chosen animals with respect, and remember that their welfare is paramount to the success of your study.

3.1 Suitable Species for Study

Selecting the right species is paramount. Your entire project hinges on this decision, influencing not only the feasibility of your research but also the quality and interpretability of the data you gather. Think of it as laying the foundation for a house; a shaky foundation will inevitably lead to structural problems down the line. Consider, first, the practical aspects: ease of rearing in a controlled environment is crucial. Can you readily obtain the species? Are its dietary needs easily met and sustainably sourced? Does it thrive in captivity, or does it require specialized and difficult-to-replicate conditions? The shorter the life cycle, generally speaking, the faster you can generate data and iterate on your experimental designs. A species with a generation time of a few weeks will allow you to conduct multiple experiments within a year, whereas a species with a generation time of several months or years will significantly slow down your research progress. Drosophila melanogaster, the common fruit fly, is a classic example. Its rapid life cycle, well-characterized genetics, and ease of maintenance make it an excellent choice for many studies. Similarly, certain species of small fish, like zebrafish (Danio rerio), are widely used due to their relatively short generation time, transparent embryos (allowing for easy observation of development), and established genetic tools. Some terrestrial organisms like Tribolium castaneum, the red flour beetle, can be easier to maintain and culture at high densities if your needs warrant.

Beyond the practicalities, delve into the biological relevance of your chosen species. Does it exhibit the traits you are interested in studying in a clear and demonstrable manner? For example, if you are investigating the effects of temperature on egg development, choose a species known to be particularly sensitive to temperature fluctuations during embryogenesis. If you're studying larval behavior, select a species with diverse and easily observable larval behaviors. Some species are naturally more prolific egg layers than others, which can be advantageous if you require a large number of eggs for your experiments. Others may exhibit more complex larval transformations, providing a richer dataset for your analyses. Always consider the specific research question you are trying to answer and choose a species that is well-suited to address that question. Moreover, it is fundamental that you are thoroughly aware of the species natural life cycle to properly evaluate the impact of each factor on them.

Ethical considerations are also non-negotiable. Your research must adhere to the highest standards of animal welfare. Before you even begin considering which species to use, familiarize yourself with the ethical guidelines and regulations governing the use of animals in research in your institution and jurisdiction. If you are working with vertebrates, for example, you will likely need to obtain approval from an animal ethics committee. This process will involve demonstrating that your research is scientifically justified, that you have taken steps to minimize any potential harm to the animals, and that you will provide them with appropriate care and housing. Even when working with invertebrates, it is important to treat them with respect and to minimize any unnecessary stress or suffering. Ensure that you are using appropriate anesthesia or euthanasia methods, if necessary, and that you are following established best practices for animal handling. Choose species that are abundant and not endangered, and prioritize those that are easily bred in captivity, reducing the need to collect individuals from the wild.

Acquisition and handling require careful planning. Once you have chosen your species, you need to obtain them from a reputable source. This could be a commercial supplier, a research institution, or, in some cases, the wild. If you are collecting animals from the wild, you must obtain the necessary permits and ensure that you are not disrupting the local ecosystem. When receiving your animals, quarantine them for a period of time to prevent the introduction of diseases or parasites into your laboratory. This is especially important if you are working with multiple species or if you are obtaining animals from different sources. Learn the proper handling techniques for your chosen species. Some species may be delicate and easily injured, while others may be capable of inflicting bites or stings. Wear appropriate protective gear, such as gloves and eye protection, and handle the animals gently and with care. Keep detailed records of your animal acquisitions, including the date of arrival, the source, and any health issues that arise. This information will be valuable for troubleshooting problems and for ensuring the long-term health and well-being of your animals. Make sure you follow the ethical guidelines to collect and handle animals.

Consider the genetic background of your chosen species. Are you working with a wild-type strain, or a genetically modified strain? If you are using a genetically modified strain, ensure that you have a clear understanding of its genotype and phenotype. The genetic background can have a significant impact on your experimental results, so it is important to control for this variable as much as possible. If possible, use a standardized strain from a reputable source. Maintain detailed records of the genetic history of your animals, including any crosses or mutations that have occurred. This information will be invaluable for interpreting your data and for replicating your experiments in the future. It is essential to know how to trace their origin for reproducibility.

Finally, remember that research is a collaborative process. Consult with experienced researchers and animal care professionals to obtain guidance on species selection and animal husbandry. Attend workshops and training sessions to learn about the latest techniques and best practices. Share your findings with the scientific community through publications and presentations. By working together, we can advance our understanding of egg laying and larval transformations while ensuring the ethical and responsible use of animals in research. Species selected will define what can be accomplished in the future.

3.2 Ethical Considerations

Ethics are at the heart of any scientific endeavor, especially when dealing with living organisms. The power to manipulate life cycles, even those as seemingly simple as egg-laying and larval development in invertebrates, carries a profound responsibility. As researchers, we must constantly question our motives, methods, and potential impact on the species we study and the wider ecosystem. This isn't just about following regulations; it's about cultivating a mindset of respect and minimizing harm in every step of our work.

When embarking on a study focused on egg-laying and larval transformations, the first ethical hurdle arises with species selection. We discussed suitable species earlier, but here, we consider the ethical implications. Wild-caught specimens, for instance, could deplete local populations or disrupt ecological balance, even if the impact is seemingly small. It's our duty to thoroughly assess the population status of any prospective species in their natural habitat. Investigate whether the species is already under pressure from habitat loss, pollution, or climate change. If so, obtaining specimens from the wild, even if legally permissible, may contribute to further decline. Could captive-bred populations provide a sustainable alternative? If so, it might be worth waiting until a captive population can produce specimens for study. If wild specimens are necessary, can collection be done sustainably, ensuring minimal impact? Perhaps collecting a few specimens from a large, healthy population, or only collecting from areas where the species is abundant and not threatened. Furthermore, consider the source of your specimens. Are they from reputable breeders who prioritize the welfare of their animals? Or are they sourced from less scrupulous dealers who may prioritize profit over ethical treatment? Transparency and traceability are key. Document the origin of all specimens and ensure compliance

with all relevant regulations.
Once you have your specimens, their care becomes paramount. Controlled environments, while essential for rigorous research, can be inherently unnatural. We must strive to create conditions that closely mimic the species' natural habitat, providing adequate space, appropriate substrate, and opportunities for natural behaviors. Enrichment is not just a buzzword; it's a moral imperative. Consider the species' natural behavior: Do they need hiding places? Opportunities to forage? Social interaction? Providing these elements can significantly improve their well-being in captivity. Crucially, we need to be vigilant in monitoring the health and well-being of our specimens. Regular observation, prompt veterinary care when needed, and humane euthanasia when suffering cannot be alleviated are all essential aspects of ethical animal care.
The manipulation of environmental parameters to influence egg-laying and larval development also raises ethical questions. While the pursuit of knowledge is valuable, we must avoid causing unnecessary distress or suffering. For example, drastically altering temperature or humidity could have detrimental effects on the organisms, leading to stress, illness, or even death. Therefore, any manipulation should be carefully justified and conducted with the utmost caution. Pilot studies can be invaluable in determining the optimal range of environmental parameters, minimizing the risk of harm to the organisms. Furthermore, ethical guidelines suggest using the minimum number of specimens needed to obtain statistically significant results. Avoid unnecessary replication and strive to maximize the informa-

tion gained from each individual.

Beyond the immediate welfare of the specimens, we must also consider the broader implications of our research. Could our findings be used to develop harmful applications? For example, could the knowledge gained about manipulating egg-laying be used to control or exploit wild populations? It is crucial to anticipate potential misuse of our research and to actively promote responsible use of scientific knowledge. Open communication and collaboration with other researchers, policymakers, and the public can help ensure that our work benefits society as a whole.

Finally, transparency and accountability are essential pillars of ethical research. Researchers need to document their methods meticulously, including the rationale for any experimental manipulations. Transparency builds trust and allows other scientists to scrutinize the work, verify the results, and identify potential ethical concerns. Disclosing funding sources and potential conflicts of interest are also important for maintaining objectivity. Similarly, all research involving living organisms should be subject to independent ethical review. Institutional Animal Care and Use Committees (IACUCs) play a crucial role in ensuring that research protocols are ethically sound and compliant with regulations. It is our shared responsibility to uphold the highest ethical standards in all our scientific endeavors, ensuring that the pursuit of knowledge is always balanced with respect for life.

Let's remember, even with the most carefully designed protocols and state-of-the-art equipment, unforeseen circumstances can arise. What happens if, despite our best efforts, the organisms suffer unexpected complications or distress? A truly ethical researcher must be prepared to halt the experiment, re-evaluate the methods, and prioritize the well-being of the specimens above all else. This requires a willingness to admit mistakes and learn from them. Ethical research is not just about following a checklist; it's about cultivating a mindset of empathy, responsibility, and continuous improvement. Moreover, it's about actively contributing to a culture of ethical awareness within the scientific community, mentoring junior researchers, and advocating for responsible research practices. The ultimate goal is to ensure that our pursuit of knowledge aligns with our commitment to protecting the planet and its inhabitants.

3.3 Acquisition and Handling

Selecting the right species for your "Phalanx Vanguard" experiments is far more than a simple choice; it's the bedrock upon which the success, ethical integrity, and scientific validity of your entire study rest. This selection must be deliberate, informed by the specific questions you aim to answer, and deeply rooted in an understanding of the organism's biology and life cycle. This chapter focuses on the ethical and practical considerations involved in acquiring and handling your chosen species, setting the stage for rigorous and humane research. Before you order your first culture or collect your first specimen, reflect deeply on the ethical implications. Are you truly minimizing harm? Can you justify the use of this particular species over alternatives? Can you provide a living environment that adequately meets their needs? Consider the species' role in its natural ecosystem. Removing individuals, especially in large numbers, could have unforeseen consequences. The best choice is often a species already well-established in laboratory settings, with readily available husbandry protocols and known sensitivity to environmental changes.

Acquisition methods will vary greatly depending on your chosen species. For commercially available organisms, such as certain beetle species or fruit flies, reputable suppliers are crucial. Ensure they maintain healthy, genetically diverse stocks and adhere to ethical collection practices. Request information on their breeding protocols and health records to minimize the risk of introducing diseases or genetic abnormalities into your controlled environment. Wild-caught specimens present a different set of challenges. If your research necessitates wild collection, obtain all necessary permits from local and national authorities. Understand the regulations regarding protected species and collection quotas. Plan your collection efforts carefully to minimize disturbance to the natural habitat and the target population. Employ humane trapping methods and prioritize the collection of adult individuals over juveniles whenever possible. Consider the seasonality of egg laying and larval development in the wild when planning your collection trips, and avoid collecting during critical breeding periods. Upon arrival, quarantine all newly acquired specimens, whether purchased or wild-caught. This period, typically lasting several weeks, allows you to observe for signs of illness or parasitism and prevent the spread of disease to your established cultures. During quarantine, maintain individuals in separate, clearly labeled containers with appropriate environmental conditions and provide a consistent food source. Regularly monitor their activity levels, feeding habits, and physical appearance. Any signs of distress, lethargy, or unusual behavior should prompt immediate investigation and consultation with a veterinarian or experienced entomologist.

Handling your chosen species requires a delicate balance between minimizing stress and ensuring efficient data collection. Before handling any organism, familiarize yourself with its specific anatomy and behavior. Some species may be easily stressed by direct contact, while others may exhibit defensive behaviors, such as biting or stinging. Use appropriate tools and techniques to minimize physical contact and avoid causing injury. Fine-tipped forceps, soft brushes, or aspirators can be useful for transferring individuals between containers or for manipulating eggs and larvae. Always work in a well-lit area to improve visibility and reduce the risk of accidental damage. When transferring organisms, avoid sudden movements or jarring motions. Gently guide them into their new environment, allowing them to acclimate before resuming your observations. Maintain a consistent routine for handling and observation to minimize disruption to their natural rhythms. Prolonged or excessive handling can lead to increased stress levels, reduced egg laying, and impaired larval development. Schedule your observations strategically to minimize the need for frequent handling and prioritize non-invasive methods whenever possible. Photography, video recording, and remote monitoring systems can provide valuable data without directly disturbing the organisms.

Beyond the immediate handling procedures, consider the long-term health and well-being of your cultures. Maintain a clean and hygienic environment to prevent the spread of disease. Regularly clean and disinfect all enclosures, tools, and work surfaces. Replace food and water frequently, and remove any waste products promptly. Monitor environmental conditions closely and make adjustments as needed to maintain optimal temperature, humidity, and light levels. Pay close attention to the diet of your chosen species. Provide a balanced and nutritious food source that meets their specific needs. Consult with experts or review published literature to determine the optimal dietary composition for egg laying and larval development. Supplementation with vitamins, minerals, or other nutrients may be necessary to ensure optimal health and reproductive success.

Finally, responsible disposal is an integral part of ethical research. At the conclusion of your experiments, humane euthanasia methods must be employed. Choose a method that is appropriate for your species and minimizes suffering. Common methods include freezing, chemical euthanasia, or physical methods such as decapitation. Consult with your institution's animal care and use committee (IACUC) to ensure that your euthanasia protocols comply with ethical guidelines and regulations. Never release experimental organisms into the wild. This can introduce invasive species, disrupt local ecosystems, and potentially transmit diseases to native populations. Properly dispose of all carcasses and waste materials according to established biohazard protocols. Thoroughly clean and disinfect all equipment and enclosures before storing them or reusing them for other purposes. By carefully considering acquisition and handling practices, you're not only prioritizing the ethical treatment of your subjects but also enhancing the quality and reliability of your research. Remember, the most valuable scientific insights are those gained through rigorous methodology and a deep respect for the living organisms at the heart of your experiments. This approach ensures that your Phalanx Vanguard efforts are both scientifically sound and ethically defensible.

4 Chapter 4: Monitoring Egg Deposition

Okay, let's talk specifics about keeping a close eye on egg-laying habits. It's not just about counting eggs; it's about understanding the "why" and "how" behind the process. This chapter dives deep into observation methods, data recording, and solving common problems that can arise in your controlled environment.

First, let's consider observation techniques. Direct visual inspection is your first line of defense, but it needs to be done carefully. You want to minimize disturbance to your subjects. Think about using a magnifying glass or a low-power microscope for closer examination of the eggs themselves – are they uniformly shaped? Are there any visible defects? Record the time of day eggs are laid. Are they crepuscular layers, diurnal or nocturnal ones? It may be species-specific. Think about setting up a camera. Time-lapse photography can be a game-changer. It allows you to monitor egg-laying activity around the clock without constantly being present, minimizing disturbance and capturing crucial details you might otherwise miss. Infrared cameras are great for night vision if your species lays eggs at night. Make sure the light from these cameras won't disturb your subjects. The placement of your observation tools is also important, consider how your actions could influence the egg-laying patterns of your species.

Next, data collection needs to be structured and consistent. Design a clear and easy-to-use data sheet, either physical or digital. Include columns for date, time of day, number of eggs laid, location within the enclosure, and any notable observations about the eggs themselves (size, color, texture). If you're studying multiple groups with different treatments, clearly label each group to avoid confusion. As you accumulate data, look for patterns. What environmental conditions correlate with increased or decreased egg production? What is the average number of eggs laid per clutch? Is there a relationship between the mother's age or diet and egg quality? Calculate relevant statistics, like the mean, median, standard deviation, and range for egg number and other relevant parameters. Charts and graphs can visually represent your data, making trends easier to identify. For example, a line graph showing egg production over time can quickly reveal whether production is increasing, decreasing, or remaining stable. Consider tracking environmental data alongside egg-laying data. This might include temperature, humidity, light levels, and even barometric pressure. Use correlation analysis to see if any of these factors are related to egg production.

What if things aren't going as planned? Now we need to troubleshoot common issues. One of the most frequent is a complete absence of egg laying. Stress is a huge factor. Are your subjects comfortable in their environment? Are they getting enough food and water? Is there too much disturbance? Check your temperature and humidity levels. Many species are very sensitive to these factors, and even slight deviations can disrupt egg laying. Make sure your temperature and humidity sensors are properly calibrated. If you suspect a nutritional deficiency, adjust their diet accordingly. Consider adding supplemental vitamins or minerals. Sometimes, the problem isn't a complete lack of egg laying, but low egg production. Consider changing the substrate if you're using one. Some species have very specific preferences for the type of material they lay their eggs in. If you're using artificial nesting sites, experiment with different designs. Some species prefer enclosed spaces, while others prefer open platforms. You can also add enrichment items to the enclosure to reduce stress and stimulate natural behaviors. These might include branches, rocks, or hiding places.

How do we boost the egg production? Enhancing egg production goes beyond simply providing the basics; you need to create an optimal environment. Consider photoperiod manipulation. Many species rely on seasonal cues, like changes in day length, to trigger reproductive activity. Mimicking these changes in your controlled environment can stimulate egg production. You can also use specific supplements. Certain vitamins and minerals are particularly important for egg production, such as calcium, vitamin D3, and vitamin E. Make sure your subjects are getting enough of these nutrients. If you have multiple females in the same enclosure, ensure there's enough space and resources to minimize competition. Overcrowding can lead to stress and reduced egg production. If you're sure your subjects have no health issue, you can consider using hormone treatment. However, hormone treatments should be used with caution and only under the guidance of a veterinarian or experienced researcher. They can have significant side effects and should be reserved for cases where other methods have failed.

Prioritize health and safety at all times. Before starting any work with your subjects, familiarize yourself with any potential hazards and implement appropriate safety measures. Wear gloves when handling eggs or cleaning the enclosure. This protects you from potential allergens or pathogens carried by the subjects. Regularly disinfect your equipment and work area to prevent the spread of disease. Use a non-toxic disinfectant that is safe for your subjects. Avoid using harsh chemicals that could harm them. When handling your subjects, be gentle and avoid stressing them unnecessarily. Use appropriate tools and techniques to minimize the risk of injury. If you are working with species that can bite or sting, take appropriate precautions to protect yourself. This might include wearing protective clothing or using specialized handling equipment. Finally, have a plan in place for dealing with emergencies, such as a power outage or a sudden illness in your subjects. This should include backup power sources, emergency contact information for a veterinarian, and procedures for isolating sick animals.

Remember, monitoring egg deposition is an ongoing process that requires patience, attention to detail, and a willingness to adapt your techniques as needed. By carefully observing your subjects, diligently recording your data, and proactively addressing any problems that arise, you can gain valuable insights into the fascinating world of egg laying and larval transformations.

4.1 Observation Techniques

Observation is at the heart of understanding egg deposition and larval transformation. Without keen observation, data collection becomes haphazard, insights are superficial, and the entire experimental endeavor risks becoming a futile exercise. It's not merely looking at your subjects; it's seeing, interpreting, and recording with purpose and precision.

Direct visual observation, despite its apparent simplicity, is a cornerstone of any study. Regular, scheduled checks of your controlled environment are essential. This might involve peering into enclosures multiple times a day, carefully noting the presence of eggs, their location, and any unusual behaviors of the egg-laying adults. Think of yourself as a silent, patient observer, blending into the background as much as possible to avoid disrupting natural behaviors. Don't just count eggs; note their distribution. Are they clustered in a specific area? Are they attached to a particular substrate? Such details can provide valuable clues about preferred laying sites and environmental preferences. The key here is consistency; establish a routine and stick to it, minimizing variations in your observation schedule.

Microscopic observation takes your analysis to a deeper level. A dissecting microscope is indispensable for examining eggs in detail. You can assess their size, shape, and color, all of which can be indicative of egg quality and health. Look for any abnormalities, such as deformities or discoloration, which might suggest underlying problems. Consider using a calibrated eyepiece reticle to make precise measurements of egg dimensions. Furthermore, microscopic examination allows you to observe the very early stages of larval development within the egg, providing insights into the initial developmental processes.

Video recording offers a powerful way to capture behavioral nuances that might be missed during brief observation periods. Set up cameras to continuously record activity within the enclosures, or use motion-activated cameras to capture specific events. Time-lapse recording can be particularly valuable for documenting egg-laying behavior and larval transformations over extended periods. When reviewing video footage, pay attention to subtle cues, such as pre-laying behaviors, interactions between individuals, and larval feeding patterns. Video analysis software can assist in quantifying these observations, allowing you to measure parameters such as activity levels, interaction frequencies, and developmental rates.

Environmental sensors provide an objective measure of the conditions within your controlled environment. Temperature, humidity, and light levels can all have a significant impact on egg laying and larval development. Deploy sensors throughout the enclosures to continuously monitor these parameters, logging the data for later analysis. Look for any correlations between environmental fluctuations and changes in egg-laying rates or larval development. For example, a sudden drop in temperature might trigger a cessation of egg laying, or a change in humidity might affect larval growth rates. Analyzing these data can help you fine-tune your controlled environment to optimize conditions for your study organisms.

Non-invasive techniques should be used whenever possible, prioritizing the well-being of your study organisms. For instance, instead of directly handling eggs for examination, consider using photography or videography to capture images and measurements. When observing larval behavior, use remote viewing systems, such as webcams, to minimize disturbance. Avoid unnecessary handling, and always prioritize creating a stress-free environment. Ethical considerations are paramount.

Data logging is as crucial as the observations themselves. Develop a standardized data sheet or electronic database to record all of your observations. Include fields for date, time, species, location, egg number, egg size, larval stage, environmental parameters, and any other relevant information. Be consistent in your data recording, and use clear, unambiguous terminology. Regularly back up your data to prevent loss, and implement quality control measures to ensure accuracy and completeness.

Troubleshooting observation challenges is often necessary. Sometimes, visibility within the enclosures can be limited, especially in complex habitats. Use supplemental lighting, such as LED lamps, to improve visibility. However, be mindful of the potential impact of artificial light on the behavior of your study organisms. If condensation obscures your view, consider using dehumidifiers or increasing ventilation. If you're struggling to identify specific larval stages, consult taxonomic keys or seek assistance from experts. Never hesitate to seek help when you encounter problems, as this can save you time and effort in the long run.

Consider the limitations of each observation technique. Direct visual observation is subjective and prone to bias. Microscopic observation can be time-consuming and may require specialized training. Video recording generates large amounts of data that can be difficult to analyze. Environmental sensors provide objective measurements, but they may not capture all of the relevant environmental variables. Therefore, a multi-faceted approach, combining different observation techniques, is often the best strategy.

Finally, remember that observation is an ongoing process, and it requires a willingness to adapt and refine your techniques as you learn more about your study organisms. Be open to unexpected findings, and don't be afraid to challenge your assumptions. The most significant discoveries often come from unexpected observations. Stay curious, stay vigilant, and let the organisms themselves guide your investigation. Ultimately, the success of your research hinges on the quality and depth of your observations.

4.2 Data Recording and Analysis

Data recording is the backbone of any scientific endeavor, particularly when dealing with the intricate processes of egg-laying and larval transformations. Without meticulous and well-organized data, our observations remain anecdotal and difficult to interpret, leading to unreliable conclusions. We need to go beyond simply noting when eggs are laid; we need a system for quantifying and qualifying the data, allowing for meaningful analysis and comparative studies.

The first step is designing a data collection sheet, whether physical or digital. This sheet should be structured to capture all relevant information about the egg-laying process. Date and time of deposition are crucial, providing a chronological record that can be correlated with environmental factors or experimental manipulations. The species and individual identification number of the egg-laying female should be clearly recorded to track individual variations and potential genetic influences. We should also note the location of the eggs within the enclosure, as this can indicate preferences or environmental gradients that influence laying behavior. The number of eggs laid in each clutch is a fundamental metric, but we should also consider other characteristics, such as egg size, shape, color, and texture. Deviations from the norm in these features can signal health issues or environmental stress. Finally, details about the substrate or oviposition site should be documented, including its composition, moisture level, and any other relevant attributes. Make sure to fill the form with accuracy and consistency.

Following egg deposition, the focus shifts to monitoring larval development. Here, the data points become more complex. Hatching rate, the percentage of eggs that successfully hatch, is a key indicator of egg viability. Upon hatching, larval size and weight should be recorded at regular intervals to track growth rates. Morphological changes, such as the number of molts or the appearance of specific anatomical features, should be carefully documented, ideally with photographic or microscopic evidence. Behavior observations are equally important. Note the larvae's feeding habits, activity levels, and responses to stimuli. Any signs of disease or abnormal development should be meticulously recorded. We must maintain detailed records of the larval diet, including the type and quantity of food provided. Document the environmental conditions throughout the larval stage, including temperature, humidity, light cycle, and water quality (if applicable). All these factors can have a significant impact on larval development, and their accurate recording is essential for interpreting any observed variations. It's a big responsibility, so be attentive.

Data analysis can range from simple descriptive statistics to more complex inferential methods. Begin with calculating basic metrics like mean, median, standard deviation, and range for egg number, egg size, hatch rate, larval growth rate, and other relevant parameters. These descriptive statistics provide a snapshot of the overall performance of your study population. Visualizing the data through graphs and charts can reveal patterns and trends that might not be apparent from raw numbers alone. Scatter plots can be used to examine correlations between variables, such as temperature and egg number, while bar graphs can compare the performance of different treatment groups. Time-series plots can track changes in larval size or weight over time. For comparing the performance of different experimental groups, statistical tests such as t-tests,

ANOVA, and chi-square tests can be used. T-tests are appropriate for comparing the means of two groups, while ANOVA can handle comparisons involving more than two groups. Chi-square tests are used to analyze categorical data, such as the proportion of larvae that survive to a certain stage. The choice of statistical test depends on the type of data being analyzed and the research question being addressed. It is important to choose correctly.

Several common pitfalls can undermine the accuracy and reliability of data collection and analysis. One frequent error is inconsistent data recording. If different observers are involved, ensure they are thoroughly trained and use standardized protocols to minimize subjective bias. Another pitfall is insufficient sample size. Small sample sizes can lead to statistically insignificant results and limit the generalizability of your findings. A third common issue is neglecting to account for confounding variables. If environmental conditions are not carefully controlled, variations in temperature, humidity, or light can obscure the effects of your experimental manipulation. Regular calibration of measurement instruments is critical. Over time, scales, thermometers, and other instruments can drift out of calibration, leading to inaccurate readings. Always use appropriate controls in your experiments. Controls provide a baseline against which to compare the effects of your experimental treatments. A control group should be treated identically to the experimental groups, except for the variable being manipulated. A final common mistake is misinterpreting statistical significance. A statistically significant result does not necessarily imply biological significance. It is important to consider the magnitude of the effect and its relevance to the biological system being studied. Take note of the limitations of each method and instrument.

Data analysis should be conducted with a clear research question in mind. Do not simply crunch numbers without a specific purpose. Formulate hypotheses about the relationships between variables and use data analysis to test these hypotheses. For example, you might hypothesize that increasing the temperature will increase the egg-laying rate of a particular species. Your data analysis should then be designed to determine whether the data support or refute this hypothesis. It is also essential to consider the limitations of your data and the potential for bias. Be honest about the uncertainties in your findings and avoid overstating the conclusions that can be drawn from your data. Transparency and rigor are crucial for maintaining scientific integrity. When presenting your data, use clear and concise language. Avoid jargon and explain statistical concepts in a way that is accessible to a broad audience. Use tables and figures effectively to summarize your findings and make them easy to understand. Remember that the goal of data analysis is not simply to generate numbers, but to gain insights into the biological processes you are studying and communicate those insights effectively to others. We are all here to explore together.

I'd like to highlight some crucial points about data security and integrity. Protecting your data from loss, corruption, or unauthorized access is paramount. Regular backups should be performed to ensure that your data can be recovered in the event of a system failure or other disaster. Data encryption can be used to protect sensitive information from unauthorized access. Restricting access to your data to only those who need it is also essential. Careful attention should be paid to data quality control. Regularly check your data for errors or inconsistencies. Implement procedures for verifying the accuracy of your data and correcting any mistakes that are found. Data integrity can be compromised by accidental or intentional alterations. Implement measures to prevent unauthorized modifications to your data. Use version control systems to track changes to your data and allow for the restoration of previous versions if necessary. It all matters, be cautious and rigorous.

Lastly, let's discuss sharing and archiving your data. Making your data publicly available allows other researchers to verify your findings and build upon your work. This promotes transparency and accelerates scientific progress. Data repositories, such as Dryad or Zenodo, provide platforms for sharing your data in a standardized and accessible format. Before sharing your data, be sure to anonymize it to protect the privacy of any individuals who may be involved. Remove any identifying information, such as names or addresses. Archive your data in a secure and durable format. Use file formats that are widely supported and unlikely to become obsolete. Store your data in multiple locations to protect against data loss. Consider using a cloud-based storage service to ensure the long-term availability of your data. Always document your data thoroughly. Provide detailed descriptions of your data collection methods, experimental procedures, and data analysis techniques. This will help others understand and interpret your data correctly. We are responsible for the quality of our work.

4.3 Troubleshooting Laying Issues

Sometimes, despite our best efforts, the carefully orchestrated egg-laying environment doesn't yield the expected results. Instead of a vibrant clutch of eggs, we might find empty enclosures, deformed eggs, or a complete cessation of laying. Before panic sets in, remember that troubleshooting is a systematic process. It begins with meticulous observation and a willingness to re-examine every aspect of our setup.

Start with the basics: temperature, humidity, and lighting. Are these parameters within the species' optimal range? A slight deviation can significantly impact egg production. Double-check your thermometers and hygrometers for accuracy. Recalibrate them if necessary, or even better, use a second set of calibrated instruments for cross-verification. Don't just rely on digital readouts; sometimes, a physical check using an old-fashioned analog thermometer can reveal subtle discrepancies. Furthermore, consider microclimates within the enclosure. Is there adequate air circulation to prevent stagnant, humid pockets? Is the light source providing the correct spectrum and intensity, or is it causing overheating in certain areas? A seemingly insignificant detail, like the placement of a heat lamp, could be the culprit. Consider investing in a data logger that can record temperature and humidity fluctuations over time, providing a more comprehensive picture of the environmental conditions.

Next, scrutinize the diet. Is the food fresh, nutritious, and readily accessible? Nutritional deficiencies are a common cause of laying problems. Ensure the food is stored properly to prevent spoilage and loss of essential vitamins. Supplement the diet with appropriate vitamins and minerals, especially if you're relying on commercially prepared feeds. However, be mindful of over-supplementation, which can also be detrimental. Observe the animals closely during feeding. Are they consuming enough food? Are there any signs of illness or parasites that might be affecting their appetite? Consider offering a variety of food items to cater to individual preferences and ensure a balanced intake. For example, some species may require live food during the breeding season to stimulate egg production. You may also need to consider the age of the food. Some ingredients degrade rapidly over time, thus the food source you are using could be old or improper.

Water is also vital for good health. Provide a clean, fresh water source at all times. Some species prefer still water, while others require dripping water or a shallow dish. Observe the animals to determine their preference. Check the water source daily for contamination or algae growth. Dehydration can significantly impact egg production and overall health. It is therefore highly important to be very careful and aware of their needs.

Stress can be a major inhibitor of egg laying. Loud noises, sudden movements, and overcrowding can all contribute to stress. Minimize disturbances in the animals' environment. Provide ample hiding places and visual barriers to create a sense of security. If you have multiple animals in the enclosure, ensure there is adequate space to avoid competition and aggression. Observe the animals' behavior closely for signs of stress, such as hiding, pacing, or aggression. Adjust the environment accordingly to reduce stressors. Consider the location of the enclosure itself. Is it in a high-traffic area or near a source of loud noise? Moving the enclosure to a quieter location may be necessary. Another common issue is the presence of potential predators, such as pets, that could be frightening your subject animals. The substrate you are using could be harmful for the animals. It may seem harmless, but it might be harboring contaminants, or it could be too damp/dry for the animal you are studying, especially if it is a soil type substrate. In some cases, especially when monitoring for disease transmission, or the presence of external pests and parasites, it may be a good idea to remove the substrate completely and observe the animals daily in an easy to clean environment.

Parasites and diseases can also disrupt egg laying. Regularly check the animals for signs of infestation, such as mites, ticks, or internal parasites. Consult with a veterinarian experienced in reptile or amphibian care for diagnosis and treatment. Implement strict quarantine protocols for new arrivals to prevent the introduction of disease into your existing colony. Practice good hygiene by regularly cleaning and disinfecting the enclosure. Use disposable gloves and equipment when handling the animals to minimize the risk of cross-contamination. Consider submitting fecal samples for analysis to detect internal parasites early.

Sometimes, the issue may be related to the breeding cycle itself. Is it the correct time of year for the species to be laying eggs? Some species have specific breeding seasons triggered by changes in temperature, light, or rainfall. Mimic these environmental cues in your controlled environment to stimulate breeding. If you are working with a seasonal breeder, ensure you are providing the correct photoperiod and temperature regime to induce reproductive activity. Research the species' natural breeding cycle thoroughly to understand their specific requirements.

If you have been keeping animals together for a long time, and the eggs are not fertile, sometimes a change in the males is the only thing you need. Especially if a male becomes too old, or has health issues, sometimes their sperm count decreases. If possible, try and introduce new males in the environment, maybe they can get things going again. You should also ensure that the ratio of males to females is correct in the setup.

Finally, sometimes it's a matter of patience. Some animals take time to adjust to a new environment before they begin laying eggs. Continue to provide optimal conditions and monitor the animals closely. Avoid making drastic changes to the environment, as this can further stress them. Document all your observations and interventions to track progress and identify potential patterns. Remember, persistence and a systematic approach are key to troubleshooting egg-laying issues.

4.4 Enhancing Egg Production

So, your organisms aren't laying as many eggs as you need for your research? Don't panic. There are several levers you can adjust within your controlled environment to coax them into boosting production. Remember, every species has its unique needs, and what works for one may not work for another. Meticulous record-keeping is absolutely essential to identify the factors that contribute most significantly to egg production in your specific study species. This is not just about getting more eggs; it's about understanding the underlying biological drivers.

First, let's think about nutrition. Diet is a cornerstone of reproductive success for pretty much any organism that lays eggs. Just like a human athlete needs specific nutrients to perform at their peak, your study subjects need a balanced diet rich in the building blocks for eggs. Consider this: the egg itself is a package of protein, fats, vitamins, and minerals designed to support the developing embryo. So, if your creatures aren't getting enough of these components in their diet, egg production will inevitably suffer. This may require experimenting with different food sources or supplements. For insects, this might mean tweaking the composition of their artificial diet or providing access to different host plants. For aquatic invertebrates, you may need to adjust the algae concentration or introduce specific micronutrients. Document every dietary change and the resulting impact on egg yield. Also, never introduce a change without proper control groups, because the results can be altered by external factors and you may be making a mistake that can put the whole experiment at risk.

Next, light and temperature are crucial environmental factors. Light cycles, or photoperiods, play a significant role in regulating the reproductive cycles of many species. Manipulation of the light-dark cycle can sometimes stimulate egg laying, especially in species with seasonal breeding patterns. Research what light cycle mimics the natural environment of your species and test those conditions. However, changing from one light cycle to another must be done very slowly, gradually increasing the hours of light instead of suddenly going from zero to twenty-four hours. Temperature is another major factor. Most organisms have an optimal temperature range for egg laying, and deviating from this range can suppress reproduction. If you are using a temperature that is too low or too high, your organisms may prioritize survival over reproduction. Again, thorough research and careful experimentation are key to discovering the ideal temperature for your species.

Now, let's get into the less obvious, but equally important, environmental variables. The humidity of the environment can affect the organism, in the eggshell deposition, and in other secondary biological functions. If humidity is too low, the eggs may dry out before they hatch, and if it is too high, the eggs may be susceptible to fungal growth. The need to maintain a stable humidity, avoiding peaks during the day or the night is essential for good egg production, as it is to minimize the amount of external stress.

Let's think about population density. Overcrowding can create stress, which then reduces egg production. Ensuring that your organisms have enough space can be crucial, and this may require adjusting the size of the enclosure or reducing the number of individuals per enclosure. On the other hand, if you have a social species, there might be a minimum number of individuals required to stimulate reproduction. So, it's not always as simple as giving them more space; you have to consider the social dynamics of your species.

Also, think about the substrate. Some species require a specific type of substrate to lay their eggs. This could be soil, sand, leaves, or even water, depending on the species. If you are not providing the right substrate, your organisms may not be able to lay their eggs properly. Ensure that the substrate is clean, free from contaminants, and appropriate for the egg-laying behavior of your species.

Beyond the physical environment, consider the social environment. For some species, the presence of a mate is essential for egg laying. Ensuring that both males and females are healthy and compatible is crucial for successful reproduction. You may need to experiment with different ratios of males to females to find the optimal balance for your species. If the species is very rare and it is difficult to find a proper couple, try to isolate the factors to understand why the couple isn't working. However, avoid making too many changes at once. It's tempting to tweak everything all at once, but this makes it impossible to determine which changes are actually having an impact. Instead, change only one variable at a time and monitor the results carefully. Allow sufficient time for the organisms to adjust to the new conditions before assessing the impact on egg production. It can take several days, weeks, or even months to see a significant change in egg laying, depending on the species and the reproductive cycle.

Also, don't overlook the potential for genetic variation. Even within a single species, there can be significant differences in egg-laying ability between individuals. If you are working with a genetically diverse population, you may be able to select for individuals with higher egg-laying rates. This could involve breeding the individuals that lay the most eggs or using genetic markers to identify individuals with favorable traits. However, be aware of the ethical implications of selective breeding and ensure that you are not inadvertently selecting for other traits that could be detrimental to the health or welfare of your species.

4.5 Health and Safety Protocols

Navigating the intricacies of egg-laying and larval transformations necessitates stringent health and safety protocols. Our controlled setups, while offering unprecedented opportunities for observation, also present unique hazards that demand careful consideration. These protocols aren't merely bureaucratic hurdles; they are the cornerstone of ethical research, ensuring the well-being of both the researchers and the organisms under study.

Implementing a robust risk assessment is the initial and most critical step. Before even introducing your chosen species into the controlled environment, meticulously identify potential hazards. Consider allergenic reactions to the organisms, particularly to their secretions or shed skins. Bites or stings, even from seemingly harmless species, can trigger unexpected reactions. Think about potential for zoonotic diseases—diseases transmissible from animals to humans—and the specific risks associated with the species you're studying. Document all identified risks, along with detailed mitigation strategies, such as personal protective equipment (PPE), specific handling procedures, and emergency contact information. Regularly review and update this assessment as your research progresses, adapting to new findings and challenges. Never underestimate the importance of this assessment, as it forms the bedrock of your safety plan.

Proper personal protective equipment (PPE) is non-negotiable. The specific PPE required will vary depending on the species you are working with and the nature of the experiment, but some basics apply across the board. Gloves are essential for handling specimens, cleaning enclosures, and working with potentially contaminated materials. Nitrile gloves are generally preferred over latex, which can trigger allergies in some individuals. Eye protection, such as safety glasses or goggles, shields your eyes from splashes, debris, and accidental exposure to potentially irritating or harmful substances. A lab coat provides a barrier between your skin and potential contaminants, preventing the spread of microorganisms and protecting your clothing. Consider respirators or masks if you're working with volatile chemicals, allergenic substances, or species that produce airborne particles. Always ensure that your PPE fits properly, is in good condition, and is used consistently throughout your work. Disposable PPE should be discarded appropriately after each use, while reusable items should be thoroughly cleaned and disinfected according to established protocols.

Safe handling procedures are paramount to minimize stress on the organisms and prevent injury to yourself.

When handling eggs or larvae, use gentle, precise movements to avoid damage. Employ tools like soft-bristled brushes, blunt forceps, or transfer pipettes to manipulate delicate specimens. Avoid direct contact with your bare hands whenever possible. If handling adult organisms, familiarize yourself with their natural defenses. Some species may bite, sting, or release irritating substances. Learn the proper restraint techniques for your chosen species to minimize stress and the risk of injury. Work in a well-lit area to ensure clear visibility and avoid accidental missteps. Maintain a calm and focused demeanor, as sudden movements or loud noises can startle the organisms and increase the likelihood of defensive behaviors. Never rush the handling process; take your time and prioritize the safety of both yourself and the animals.

Chemical safety protocols are indispensable when using disinfectants, cleaning agents, or preservatives. Always read and understand the safety data sheets (SDS) for all chemicals you use. These sheets provide comprehensive information on the hazards associated with each substance, as well as proper handling, storage, and disposal procedures. Store chemicals in designated areas, away from incompatible substances and in accordance with manufacturer's instructions. Use appropriate ventilation when working with volatile chemicals to minimize inhalation exposure. Never mix chemicals unless you are certain that the combination is safe and will not produce hazardous fumes or reactions. Wear appropriate PPE, such as gloves, eye protection, and a respirator, when handling chemicals. In the event of a spill, clean it up immediately using appropriate spill control materials and following established protocols. Dispose of chemical waste properly, in accordance with local regulations.

Waste management protocols are critical for preventing the spread of disease and maintaining a clean, safe environment. Separate different types of waste, such as general trash, biological waste, and chemical waste, into designated containers. Biological waste, including dead organisms, eggshells, and contaminated materials, should be autoclaved or otherwise sterilized before disposal. Sharps, such as needles, scalpel blades, and broken glass, should be disposed of in puncture-resistant sharps containers. Chemical waste should be disposed of in accordance with local regulations and the manufacturer's instructions. Regularly empty waste containers to prevent overflow and odor buildup. Clean and disinfect waste collection areas to minimize the risk of contamination. Maintain detailed records of waste disposal activities, including the type and quantity of waste, the disposal method, and the date of disposal.

Emergency procedures must be clearly defined and readily accessible. Develop a comprehensive emergency plan that outlines the steps to take in the event of an accident, injury, or other emergency. This plan should include contact information for emergency services, first aid providers, and relevant personnel. Post emergency contact information prominently in the laboratory. Train all personnel in emergency procedures, including first aid, CPR, and the use of safety equipment. Conduct regular drills to ensure that everyone is familiar with the emergency plan and can respond effectively in a crisis. Maintain a well-stocked first aid kit in a readily accessible location. Report all accidents and injuries, no matter how minor, to the appropriate authorities. Investigate all incidents to identify the root causes and implement corrective actions to prevent future occurrences.

Hygiene practices are essential for preventing the spread of contamination. Wash your hands thoroughly with soap and water before and after handling organisms, cleaning enclosures, or working in the laboratory. Avoid touching your face, eyes, or mouth while working. Keep your work area clean and organized. Regularly disinfect surfaces and equipment with appropriate cleaning agents. Change your gloves frequently, especially after handling different organisms or materials. Avoid eating, drinking, or smoking in the laboratory. Store food and drinks in designated areas, away from potentially contaminated materials.

Training and documentation form the backbone of a successful health and safety program. All personnel working with egg-laying organisms and larvae must receive comprehensive training on safety protocols, handling procedures, and emergency response. This training should be documented and updated regularly. Maintain detailed records of all experiments, including species information, environmental conditions, procedures, and any deviations from established protocols. This documentation is essential for troubleshooting problems, replicating results, and ensuring the reproducibility of your research. Regular audits of safety protocols and procedures should be conducted to identify areas for improvement and ensure compliance with regulations.

By meticulously implementing these health and safety protocols, we not only safeguard ourselves and the organisms under our care but also ensure the integrity and reliability of our research. These practices are not just about avoiding accidents; they are about fostering a culture of responsibility and respect within the laboratory, contributing to a more ethical and sustainable approach to scientific inquiry.

5 Chapter 5: Larval Stage Transformations

Larval development is a dynamic period marked by dramatic changes in form and function. These transformations are not merely passive processes dictated solely by genetic programming; they are highly sensitive to environmental cues and nutritional inputs. Understanding these interactions is crucial for a comprehensive study of the life cycle in our controlled environment. We can break this down into understanding the process, providing appropriate nutrition, and manipulating the environment.

Nutrition plays a pivotal role in dictating the success of larval development. The nutritional requirements of larvae can vary significantly based on the species you are studying. For instance, some insect larvae are highly specific in their dietary needs, requiring particular host plants or prey items to thrive. Others are more generalists, capable of utilizing a wider range of food sources. Therefore, a thorough understanding of the natural diet of your chosen species is essential. This understanding should inform the formulation of an artificial diet that closely mimics the nutritional profile of their natural food. This can involve sourcing specific ingredients, blending food items, or even synthesizing essential compounds. This is also when you have to think about the future. What form will they take? How will it affect them? The answer usually lies in the nutrition.

Environmental conditions also have a profound impact on larval development. Temperature, humidity, photoperiod (light/dark cycle), and substrate availability can all influence growth rate, molting frequency, and ultimately, the size and form of the emerging adult. Manipulation of these environmental factors offers powerful opportunities to investigate the plasticity of development and to uncover the underlying mechanisms that regulate these processes. For example, by rearing larvae at different temperatures, you can examine the effects of temperature on developmental time and body size. Similarly, varying the photoperiod can influence the timing of metamorphosis and the expression of specific genes involved in development. The larval stage is a period of intense vulnerability.

Observation of larval behavior offers a non-invasive window into their physiological state and developmental progress. Monitoring feeding rate, activity level, and social interactions (if applicable) can provide valuable insights into their overall health and well-being. Moreover, these observations can be correlated with other data, such as growth rate and molting frequency, to provide a more holistic understanding of their development. It is critical to document all this as objectively as possible to get true measurements of larval development.

One effective strategy for enhancing larval growth is to optimize the rearing environment. This may involve providing a carefully controlled temperature gradient, allowing larvae to select their preferred microclimate. Similarly, adjusting humidity levels can promote healthy molting and prevent desiccation. Providing an appropriate substrate, such as soil, leaf litter, or artificial nesting materials, can also encourage natural behaviors, such as burrowing, feeding, and pupation. These factors can be altered or enhanced by creating the optimal environment for the specific needs of the insect you are studying.

To ensure accurate and reliable data, it is essential to establish standardized protocols for measuring larval size and weight. This may involve using calipers, micrometers, or digital scales to obtain precise measurements. Regular measurements should be taken at fixed intervals, such as daily or weekly, to track growth over time. It is also important to document any abnormalities or deviations from the norm, such as stunted growth, deformities, or unusual coloration. The more data we have, the better we can track larval growth.

Beyond growth, you can dive deeper into the underlying mechanisms governing larval development by examining gene expression patterns. Techniques such as quantitative PCR (qPCR) or RNA sequencing can be used to measure the abundance of specific mRNA transcripts, providing insights into the activity of key genes involved in development. These techniques can be particularly useful for identifying genes that are differentially expressed in response to environmental manipulations or nutritional interventions. By combining gene expression data with phenotypic observations, you can gain a more comprehensive understanding of the molecular basis of larval development. It is through these methods that we can uncover the hidden mechanisms that drive change.

Ethical considerations are paramount when working with live organisms. It is essential to minimize stress and suffering and to ensure that the animals are treated humanely. This includes providing appropriate housing, nutrition, and environmental conditions, as well as avoiding unnecessary handling or experimentation. In cases where euthanasia is necessary, it should be performed using humane methods that minimize pain and distress. Ethical guidelines and regulations should be strictly adhered to at all times. With ethical consideration, you can minimize the harm your studies could potentially cause.

The larval stage represents a critical window for understanding the interplay between genes, environment, and nutrition. By carefully controlling and manipulating these factors, we can gain valuable insights into the developmental processes that shape the lives of organisms. These insights can have broad implications for fields such as agriculture, medicine, and conservation. The importance of what we can learn at this stage is paramount to the health of not only the organisms we study, but other organisms as well.

Remember to meticulously document all procedures, observations, and data. This includes detailed records of environmental conditions, feeding schedules, growth measurements, and any other relevant information. Thorough documentation is essential for ensuring the reproducibility of your experiments and for sharing your findings with the broader scientific community. This also allows for reflection and changes if you feel that is needed.

Be prepared to adapt your experimental design as needed. Larval development is a complex process, and unexpected challenges may arise. Be flexible and willing to modify your protocols in response to new information or observations. The ability to adapt is a key skill for any successful researcher. It is through this adaptation, we can overcome the challenges of studying this period.

5.1 Overview of Larval Development

Larval development represents a critical phase in the life cycle of many insects and invertebrates, bridging the embryonic stage with adulthood. This period is characterized by intense feeding, rapid growth, and significant morphological changes, making it a fascinating area of study for developmental biologists and entomologists alike. Understanding the intricacies of larval development, including the factors that influence it, is essential for anyone working with these organisms in controlled environments.

The most obvious hallmark of larval development is growth. Newly hatched larvae are typically small and vulnerable, possessing limited mobility and feeding capabilities. As they consume food, they increase in size and biomass, accumulating the resources needed for subsequent metamorphosis. This growth is not continuous; instead, it occurs in discrete stages called instars, separated by molts. During each molt, the larva sheds its old exoskeleton and emerges with a larger, more flexible one. The number of instars varies depending on the species and environmental conditions. Nutrition plays a pivotal role during this time; adequate food supply and a balanced diet are crucial for healthy growth and successful metamorphosis. Deficiencies can lead to stunted growth, developmental delays, or even mortality. The specific nutritional requirements of larvae vary widely, depending on their feeding habits and the type of food they consume. Some larvae are generalists, feeding on a wide range of plant or animal matter, while others are specialists, requiring specific nutrients from a limited number of food sources.

Beyond simple growth, larval development involves significant morphological transformations. Many larvae look dramatically different from the adults they will eventually become. This is particularly true in insects with complete metamorphosis (holometabolism), such as butterflies, beetles, and flies. These insects have a distinct larval stage that is specialized for feeding and growth, followed by a pupal stage during which the adult body plan is assembled. In contrast, insects with incomplete metamorphosis (hemimetabolism), such as grasshoppers and dragonflies, have nymphs that resemble smaller versions of the adults and gradually develop adult features through successive molts. The morphological changes that occur during larval development are controlled by complex genetic and hormonal mechanisms. Specific genes are activated and deactivated at different stages of development, leading to the differentiation of cells and tissues. Hormones, such as ecdysone and juvenile hormone, play a crucial role in regulating molting and metamorphosis. The relative levels of these hormones determine whether a larva will molt into another larval instar or undergo metamorphosis into a pupa or adult.

Environmental factors exert a powerful influence on larval development. Temperature is a critical factor, as it affects the rate of metabolism and development. Larvae typically develop faster at higher temperatures, but only within a certain range. Temperatures that are too high or too low can be lethal. Humidity is also important, as it affects the rate of water loss from the larva's body. Larvae that live in dry environments need to be able to conserve water, while those that live in moist environments are less susceptible to desiccation. Photoperiod, or the length of day and night, can also influence larval development. Some larvae require specific photoperiod cues to initiate metamorphosis. For instance, some insects enter diapause, a state of dormancy, during the winter months in response to short day lengths. Population density can also affect larval development. High population densities can lead to increased competition for food and resources, which can result in slower growth rates and smaller adult sizes.

Studying larval development in a controlled environment offers several advantages. It allows researchers to isolate the effects of specific environmental factors and manipulate them to investigate their impact on larval growth and metamorphosis. For example, one can precisely control the temperature and humidity to determine the optimal conditions for larval development. Similarly, one can manipulate the diet to study the effects of different nutrients on larval growth and metamorphosis. Controlled environments also make it easier to monitor larval development closely. Researchers can track the growth rate, molting frequency, and morphological changes of individual larvae over time. This allows for a more detailed understanding of the developmental process.

Understanding the processes associated with larval development also provides key insights into ecological interactions. For example, the timing of larval development can be critical for synchronizing with the availability of food resources or avoiding predators. The morphological adaptations of larvae can also influence their interactions with their environment. For instance, larvae with camouflage coloration are better able to avoid detection by predators, while larvae with specialized mouthparts are better able to exploit specific food sources. Studying the interactions of larval development with environmental pressures is crucial to understanding evolutionary processes in animals, thus connecting developmental biology and zoology.

In summary, larval development is a dynamic and complex process that is influenced by a variety of factors, including genetics, nutrition, and the environment. By understanding the intricacies of larval development, one can gain valuable insights into the biology of insects and invertebrates. This knowledge is essential for anyone working with these organisms in controlled environments, as it allows for the optimization of rearing conditions and the manipulation of development for research purposes. This foundational understanding paves the way for more in-depth exploration of specific aspects of larval development, such as nutritional requirements and the effects of environmental manipulation, which are crucial for successful rearing and experimentation.

5.2 Nutritional Requirements

Understanding the nutritional needs of larvae is crucial for successful rearing and experimentation. These minute organisms require specific nutrients in precise amounts to fuel their rapid growth, development, and metamorphosis. What constitutes an adequate larval diet varies significantly depending on the species in question, as their evolutionary adaptations have led to diverse feeding strategies. Some larvae are voracious predators, consuming other small organisms, while others are herbivores, grazing on plant matter, and still others are detritivores, feeding on decaying organic material. Providing the appropriate food source is paramount to their survival and subsequent developmental stages.

Formulating a suitable larval diet isn't simply about providing food; it's about ensuring the nutritional composition is optimized for the specific species. A diet lacking in essential amino acids, vitamins, or minerals can lead to stunted growth, developmental abnormalities, or even mortality. For example, some insect larvae require specific sterols, like cholesterol, which they cannot synthesize themselves and must obtain from their diet. These sterols are crucial for hormone production, which regulates molting and metamorphosis. Similarly, the ratio of carbohydrates to proteins in the diet can influence larval growth rate and body size.

Creating artificial diets can be particularly challenging but offers the greatest control over nutritional intake. These diets often consist of a complex mixture of ingredients, including purified proteins, carbohydrates, lipids, vitamins, and minerals. Determining the optimal concentration of each component requires careful experimentation, often involving the systematic alteration of the diet and monitoring of larval growth and development. For instance, one might start with a base diet and then vary the amount of protein to determine the level that maximizes larval weight gain and survival. Similarly, different sources of carbohydrates, such as glucose, fructose, or sucrose, might be tested to identify the most efficiently utilized energy source.

Preparation and delivery of larval food are also important considerations. Many larvae have delicate mouthparts and require food particles of a specific size and texture. Furthermore, the food must be presented in a way that allows the larvae easy access to it. This might involve suspending food particles in a gel, coating a surface with a thin layer of food, or providing a continuous supply of fresh food. Contamination of the food with bacteria or fungi can also be a major problem, so sterile techniques are often necessary to prevent the introduction of pathogens.

Beyond the basic nutritional components, the presence of specific additives can also enhance larval growth and development. These additives might include antioxidants, which protect against oxidative damage, or prebiotics, which promote the growth of beneficial gut bacteria. The composition of the gut microbiome can significantly influence larval health and development, as these bacteria can aid in digestion, synthesize essential nutrients, and protect against pathogens. Manipulating the gut microbiome through dietary interventions is an area of growing interest in larval rearing.

Environmental factors can also interact with nutritional intake to influence larval development. Temperature, humidity, and light levels can all affect larval metabolism and appetite. For example, larvae reared at higher temperatures may require more food to maintain their growth rate. Similarly, humidity levels can affect the rate of water loss from the larvae, which can influence their ability to absorb nutrients. Providing an optimal environment is therefore essential for maximizing the benefits of a well-formulated diet.

Observing larval behavior provides valuable insights into their nutritional status. Healthy larvae typically exhibit active feeding behavior, while malnourished larvae may be sluggish and disinterested in food. Changes in larval color, size, or shape can also indicate nutritional deficiencies. Regular monitoring of larval health and behavior can allow for early detection of problems and prompt adjustments to the diet. Careful record-keeping of larval growth rates, survival rates, and developmental milestones is essential for evaluating the effectiveness of different diets.

Supplementation is another important aspect. In some cases, even a well-formulated diet may be lacking in certain nutrients. Supplementation with specific vitamins, minerals, or amino acids can address these deficiencies and improve larval health. For example, some insect larvae require a dietary source of carotenoids to produce pigments that protect against UV damage. Supplementation with carotenoids can improve larval survival and coloration. It is important to note, however, that over-supplementation can also be harmful, so caution is advised.

Considering the long-term effects of larval nutrition is necessary. The diet consumed during the larval stage can have lasting effects on adult phenotype, including size, fecundity, and lifespan. For instance, larvae that are raised on a nutrient-rich diet may develop into larger adults with higher reproductive output. This phenomenon, known as "carry-over effects," highlights the importance of providing optimal nutrition during the larval stage to maximize the fitness of the adult organism. The legacy of larval diet can also have evolutionary implications, influencing the adaptation of populations to different environments.

Furthermore, when choosing between fresh natural food and prepared food, know that while some species thrive on readily available, commercially prepared diets, others require fresh, natural food sources to complete their development. This difference can be due to the presence of specific micronutrients or enzymes in the natural food that are absent in artificial diets. The use of fresh resources may pose logistical challenges but could be essential to the success of the experiment.

Lastly, remember that the quest for the perfect larval diet is an ongoing process. As we learn more about the nutritional requirements of different species, we can continue to refine our diets and improve larval rearing techniques. This iterative process of experimentation and optimization is essential for advancing our understanding of larval biology and for developing sustainable aquaculture and insect farming practices. This knowledge enhances our ability to study developmental biology and contributes to our understanding of ethology and evolutionary adaptations.

5.3 Environmental Manipulation

Manipulating the environment during the larval stage offers researchers a potent tool to dissect developmental processes, pushing beyond mere observation to actively shape the trajectory of an organism's growth. This controlled interference allows us to explore the limits of phenotypic plasticity, teasing apart the relative contributions of genes and environment to the final form. It's about understanding not just what is, but what could be under different selective pressures.

Temperature, perhaps the most frequently manipulated factor, exerts profound influence over larval development. Lowering the temperature, for example, can significantly slow metabolic rates, extending the larval stage and potentially leading to larger adult sizes in some species. Conversely, raising the temperature can accelerate development, resulting in smaller, faster-maturing individuals. These alterations aren't simply about speed; they can trigger changes in morphology, affecting everything from appendage size to pigmentation patterns. To achieve precise temperature control, incubators and environmental chambers are essential, allowing for degree-by-degree adjustments and consistent maintenance. It's crucial to monitor temperature within the larval enclosures themselves, as microclimates can form even within a controlled environment. Water baths can be used for aquatic species, providing a highly stable thermal environment. Detailed records of temperature fluctuations, even minor ones, should be meticulously kept, as these variations can have subtle yet cumulative effects on larval development.

Light, another critical environmental cue, plays a significant role in regulating development, particularly in species with visually guided behaviors or light-dependent physiological processes. Altering the photoperiod—the duration of light exposure—can influence developmental timing, triggering metamorphosis prematurely or delaying it. Light intensity also matters; some species require high-intensity light for proper pigment development, while others thrive in dimly lit environments. For species sensitive to specific wavelengths, specialized lamps that emit narrow bands of light can be used to target particular photoreceptors and developmental pathways. Setting up light cycles that mimic natural conditions is often a good starting point, but the real power lies in deviating from these norms to explore the boundaries of developmental plasticity. Careful consideration must be given to the type of lighting used, avoiding those that emit excessive heat, which could confound the results. Controlling light is not just about turning lamps on and off; it's about shaping the larval experience in a way that reveals the underlying mechanisms of development.

Beyond temperature and light, chemical cues in the environment can also be strategically manipulated.

Adding hormones or hormone mimics to the larval environment can induce precocious metamorphosis, allowing researchers to study the effects of hormonal signaling on specific developmental events. Introducing pollutants or toxins, in carefully controlled doses, can reveal the vulnerability of different developmental stages to environmental stressors. The possibilities are vast, ranging from the addition of specific amino acids to assess their impact on growth to the introduction of signaling molecules that influence cell differentiation. When working with chemical cues, it's essential to ensure that the chosen substance is stable in the larval environment and doesn't degrade or interact with other components in unpredictable ways. Accurate dosing is paramount, and control groups must be meticulously maintained to account for any non-specific effects of the delivery method. This is a step that you must do to achieve more success.

Furthermore, the physical environment itself can be modified to probe developmental mechanisms. Altering the substrate on which larvae develop, for instance, can affect their locomotor behavior and muscle development. Varying the water flow rate in aquatic systems can influence feeding rates and oxygen uptake. Even the presence or absence of specific structures, such as vertical surfaces for climbing, can impact larval morphology and behavior. Consider the substrate's texture: a rough surface may stimulate sensory development differently than a smooth one. Think about the availability of refuge: providing hiding places can reduce stress and influence social interactions. Manipulating the physical environment is about more than just creating a comfortable home for the larvae; it's about creating an environment that challenges them in specific ways, forcing them to adapt and revealing the limits of their developmental potential.

But with great power comes great responsibility. Environmental manipulations must always be conducted ethically and with careful consideration for the welfare of the larvae. Any manipulation that causes undue stress or suffering is unacceptable. Before undertaking any experiment, a thorough ethical review should be conducted, weighing the potential benefits of the research against the potential harm to the animals. The number of larvae used should be minimized, and the experimental conditions should be optimized to reduce stress. Regular monitoring of larval health and behavior is essential, and any signs of distress should be addressed immediately. The goal of environmental manipulation is to learn about development, not to inflict harm. It is more relevant to carefully consider the ethical implications of your study.

Ultimately, environmental manipulation is not just about changing the conditions in which larvae develop; it's about asking fundamental questions about the nature of development itself. By carefully controlling and altering the environment, we can gain insights into the complex interplay between genes, environment, and phenotype, advancing our understanding of evolution, developmental biology, and the adaptability of life in a changing world. This is what developmental biology is all about. As you get deeper into the research process, remember the power and responsibility that come with manipulating the environment of these developing organisms.

6 Chapter 6: Data Analysis and Interpretation

Data, in its raw form, is simply a collection of observations. It tells no story until we, as researchers, begin to weave a narrative around it. This chapter is your guide to transforming those raw numbers and observations into insightful conclusions about egg-laying and larval transformations. We'll move beyond simple data collection and dive into the analytical techniques that will allow you to not just see, but to truly understand the biological processes you're studying.

The first step is choosing the right statistical tools. The selection hinges on the type of data you've collected and the questions you're trying to answer. For instance, if you're comparing the average number of eggs laid under different temperature conditions, a t-test or ANOVA (Analysis of Variance) might be suitable. A t-test is appropriate when comparing the means of two groups, while ANOVA is used when comparing the means of three or more groups. It's vital to understand the assumptions of each test; for example, most parametric tests assume that your data is normally distributed. Before applying any statistical test, assess whether your data meets these assumptions. If not, consider non-parametric alternatives like the Mann-Whitney U test (for comparing two groups) or the Kruskal-Wallis test (for comparing three or more groups), which do not require normally distributed data. Choosing the wrong test can lead to misleading conclusions, so consult with a statistician if you're unsure. Furthermore, when examining larval development, consider regression analysis to understand how different environmental factors influence growth rates. Linear regression can be used to model the relationship between a continuous variable (e.g., larval size) and one or more predictor variables (e.g., temperature, food availability). Understanding which tests to use will greatly improve your analysis and allow for better understanding.

Now, let's talk about interpreting larval changes. Larval development is a dynamic process influenced by many factors, and analyzing it requires a multi-faceted approach. Size, weight, developmental stage, and behavior are all important metrics to consider. If you're studying the effects of a specific treatment on larval development, compare the growth rates and developmental timelines of treated larvae with those of a control group. Pay attention to any morphological abnormalities or behavioral changes, as these can be important indicators of the treatment's impact. In addition to quantitative data, qualitative observations can provide valuable insights. Document any unusual behaviors, feeding patterns, or physical characteristics observed in the larvae. If you're using microscopy to examine larval tissues, note any cellular or structural changes that occur in response to the treatment. It is important to be as detailed as possible. Be aware of potential confounding factors that could influence larval development, such as genetic variation, maternal effects, or variations in food quality. Try to control for these factors in your experimental design or account for them in your analysis. Consider survival rates too, as they provide vital information about the overall health and fitness of the larvae under different conditions.

Drawing meaningful conclusions requires careful consideration of your data in the context of existing knowledge. Start by summarizing your findings in a clear and concise manner. What were the main effects of your experimental treatments on egg laying and larval development? Were there any unexpected results or patterns in your data? Then, compare your results with those of previous studies. Do your findings support or contradict existing theories about egg laying and larval development? If your results differ from those of previous studies, consider possible explanations for the discrepancies. This will highlight a potential gap in the field and could lead to further investigation. Discuss the limitations of your study. Are there any factors that could have influenced your results? Are there any alternative explanations for your findings? Being transparent about the limitations of your study will enhance its credibility. Consider the broader implications of your findings. How do your results contribute to our understanding of egg laying and larval development? Do they have any practical applications in areas such as pest control or conservation biology? What are the next steps for research in this area? What questions remain unanswered? By identifying future research directions, you can help advance our knowledge of these fascinating biological processes. Remember, the goal is to use your data to tell a compelling story about egg laying and larval transformations, and to contribute to the broader scientific understanding of these phenomena.

Consider the context of your study, the environmental conditions, and any inherent biases that might skew the results. For example, if you are studying the effect of temperature on egg laying, you should consider that humidity levels could also be playing a role. This is where the statistical control of variables comes into play. By statistically controlling for these variables, you can isolate the effect of temperature on egg laying. This is not to be overlooked, it can bring a lot of noise and hinder proper analysis.

Remember, data analysis is an iterative process. You may need to revisit your methods and adjust your approach as you gain a better understanding of your data. Don't be afraid to experiment with different statistical techniques and visualization methods to find the best way to communicate your findings. Consult with colleagues, mentors, or statisticians to get feedback on your analysis and interpretation. Incorporating diverse perspectives can help you identify potential biases or oversights. Furthermore, clear and concise communication of your findings is essential for sharing your research with the broader scientific community. Use appropriate visuals, such as graphs and charts, to illustrate your data. Write a clear and well-organized report or manuscript that describes your methods, results, and conclusions. It's important to adhere to the ethical guidelines of scientific research. Ensure that your data is accurate and reliable, and avoid any form of manipulation or falsification. Always acknowledge the contributions of others and give credit where credit is due. By upholding the highest standards of scientific integrity, you can contribute to the advancement of knowledge and build trust in your research.

Finally, never underestimate the power of visualization. A well-crafted graph can often reveal patterns and relationships that are not immediately apparent from the raw data. Use scatter plots, bar graphs, line graphs, and other types of visualizations to explore your data and communicate your findings effectively. Experiment with different types of visualizations to find the ones that best illustrate your key results. Be sure to label your axes clearly, and provide a concise caption that explains the meaning of the graph.

6.1 Statistical Methods

Statistical methods form the backbone of robust scientific inquiry, providing the framework for extracting meaningful insights from the often complex and variable data collected during egg-laying and larval development studies. We are diving into some practical applications rather than theoretical statistics.

Descriptive statistics offer a crucial starting point. They provide a concise summary of your collected data. Imagine you are tracking the number of eggs laid per day across different enclosures with varying temperatures. Calculating the mean (average) number of eggs laid per day for each temperature group provides an immediate sense of the central tendency. However, the mean alone tells an incomplete story. The standard deviation, a measure of data dispersion, reveals how much the individual egg counts vary around that mean. A large standard deviation indicates considerable variability, suggesting other factors might be influencing egg laying beyond just temperature. Similarly, descriptive statistics like the median (the middle value) and mode (the most frequent value) can be valuable, particularly when dealing with non-normal data distributions. For instance, if a few enclosures experience exceptionally high egg-laying days due to unforeseen events, the median might be a more robust measure of central tendency than the mean, which can be skewed by outliers. Calculating frequencies and percentages for discrete data is also vital. If you are categorizing larvae based on morphological characteristics (e.g., presence or absence of certain features), these metrics reveal the proportion of larvae falling into each category. Graphical representations, such as histograms and boxplots, are indispensable tools for visualizing descriptive statistics. Histograms display the distribution of egg counts, allowing you to quickly assess skewness and identify potential bimodality (two peaks), which might suggest different sub-populations within your sample. Boxplots, on the other hand, provide a visual comparison of egg-laying rates across different experimental groups, highlighting the median, quartiles, and any outliers.

Moving beyond simple descriptions, inferential statistics allow us to draw conclusions about the broader population based on our sample data. T-tests are commonly used to compare the means of two groups. For example, you could use an independent samples t-test to determine whether there is a statistically significant difference in the average larval size between those raised on two different diets. The t-test calculates a t-statistic and a corresponding p-value. The p-value represents the probability of observing the data (or more extreme data) if there is truly no difference between the groups. A small p-value (typically less than 0.05) provides evidence against the null hypothesis (the hypothesis of no difference), suggesting that the observed difference is statistically significant. Analysis of Variance (ANOVA) extends this comparison to more than two groups. Imagine you are investigating the effect of three different light intensities on larval development time. ANOVA allows you to determine whether there is a significant difference in the average development time across these three light intensity groups. If ANOVA yields a significant result, it indicates that at least one of the groups differs significantly from the others. Post-hoc tests, such as Tukey's HSD or Bonferroni correction, are then used to determine which specific pairs of groups differ significantly. ANOVA assumes that the data are normally distributed and have equal variances across groups. If these assumptions are violated, non-parametric alternatives like the Kruskal-Wallis test should be considered. Correlation and regression analysis are useful for examining the relationships between variables. Correlation measures the strength and direction of a linear association between two continuous variables, for example the correlation between the size of the egg and the eventual size of the larvae. A positive correlation indicates that as one variable increases, the other tends to increase as well. Regression analysis goes a step further by allowing you to predict the value of one variable (the dependent variable) based on the value of another variable (the independent variable). For example, you might use regression to predict the number of eggs laid based on the ambient humidity.

Non-parametric tests are crucial when the assumptions of normality or equal variances are not met. The Mann-Whitney U test, for instance, serves as a non-parametric alternative to the independent samples t-test, comparing two groups when the data are not normally distributed. Similarly, the Wilcoxon signed-rank test is used for paired data when the normality assumption is violated. For comparing more than two groups, the Kruskal-Wallis test is the non-parametric counterpart to ANOVA. Chi-square tests are used to analyze categorical data. For example, you might use a chi-square test to determine whether there is a statistically significant association between the type of substrate used for egg laying (e.g., leaf litter, soil, bark) and the hatching success rate. The chi-square test compares the observed frequencies of each category with the expected frequencies under the assumption of no association. A significant chi-square statistic suggests that the observed association is unlikely to have occurred by chance.

Power analysis is an essential step in experimental design, determining the minimum sample size needed to detect a statistically significant effect of a given size. A study with insufficient power is unlikely to detect a real effect, even if one exists. Power analysis takes into account the desired significance level (alpha), the desired power (typically 80% or higher), and the estimated effect size (the magnitude of the difference or association you expect to observe). Several software packages and online tools are available to perform power analysis. Understanding and appropriately using statistical methods is vital for extracting valid and meaningful conclusions from egg-laying and larval transformation studies. Descriptive statistics provide essential summaries of the data. Inferential statistics allow you to test hypotheses and draw inferences about the broader population. Non-parametric tests offer alternatives when the assumptions of parametric tests are violated. Always remember to report effect sizes and confidence intervals alongside p-values to provide a complete picture of your findings. By mastering these statistical tools, researchers can ensure the rigor and reliability of their work, advancing our understanding of these complex biological processes.

6.2 Interpreting Larval Changes

Larval transformations present a cascade of observable changes, each providing a window into the organism's developmental processes. However, observing isn't enough. We need to decode what these changes signify, correlating them with environmental factors, genetic predispositions, and even subtle nuances in diet. It's akin to reading a biological language; each moult, color change, or behavioral shift represents a sentence within the larger narrative of development. To begin, meticulous documentation is paramount. Don't rely on memory; invest in a system for recording observations. This could be a simple notebook, a spreadsheet, or even a dedicated software program. Record the date, time, temperature, humidity, and any other relevant environmental parameters. Note the larval stage, size (length and width), color, any distinctive markings, and their behavior (feeding, movement, social interactions, if applicable). Photographs or videos are invaluable for later comparison and analysis. They provide a visual record that can be easily shared and reviewed, aiding in identifying subtle changes that might otherwise be missed. The most common error beginners make is treating all larval changes as equal. A change in size is drastically different from a change in pigmentation, and their implications for the organisms well-being and the experiment's validity vary. Therefore, let's focus specifically on some key changes. First, consider size. Gradual increases in size between molts are expected, indicating healthy growth. However, deviations from this norm can signal problems. Stunted growth could be due to insufficient nutrition, unfavorable environmental conditions, or even disease. Conversely, unusually rapid growth might suggest an overabundance of resources or a genetic anomaly. Correlate size changes with food consumption and environmental parameters to identify potential causes. Record sizes over time and then check for correlation. Next, molting is a critical process in larval development. The frequency and timing of molts are species-specific and can be influenced by environmental factors. Delayed molting might indicate nutritional deficiencies or low temperatures, while premature molting could be triggered by stress or hormonal imbalances.

Examine the shed exoskeletons for completeness and any signs of damage. A deformed or incomplete exoskeleton suggests problems with the molting process itself, possibly due to calcium deficiency or exposure to toxins.

Furthermore, examine variations in pigmentation. Larval color often changes as they develop, sometimes serving as camouflage or signaling maturity. However, abnormal color changes can be indicative of underlying issues. For example, a loss of pigmentation might suggest nutritional deficiencies or exposure to harmful chemicals. Conversely, an intensification of color could be a stress response or a sign of disease. Compare the color changes to established norms for the species and investigate any deviations. Are the larvae clustered or distributed evenly? Do they exhibit coordinated movements, or are they isolated and lethargic? Changes in social behavior can be particularly informative, as they often reflect the overall health and well-being of the larvae. Reduced activity levels might indicate illness or starvation, while increased aggression could be a sign of overcrowding or competition for resources. Observe feeding behavior closely, too. A healthy larva will typically feed actively and consistently. Reduced feeding or complete cessation of feeding is a red flag, suggesting a problem with the diet, the environment, or the larva itself. Take detailed notes on the type and amount of food consumed, as well as any signs of regurgitation or other digestive issues.

The next step involves correlating the changes you observe with the environmental parameters you're controlling. Remember that in this setup, you are the master of the environment. You decide temperature, humidity, light cycle, and food availability. Thus, every change in the larva must be considered in relation to these conditions. For instance, if you observe stunted growth, cross-reference this with your temperature records. Was there a sudden drop in temperature that might have slowed down their metabolism? Or perhaps the humidity was too low, leading to dehydration and impaired growth? Similarly, if you notice abnormal pigmentation changes, check your light cycle. Was there an unexpected change in the intensity or duration of light exposure that could have affected pigment production? Don't underestimate the importance of small changes in food quality and food origin. Maybe a supplier switch changed its formulation, and your larvae are now reacting to a new preservative or some other ingredient.

After that, genetic background can play a significant role in larval development. Even within a single species, there can be variations in growth rate, coloration, and behavior due to genetic differences. If possible, obtain information about the genetic lineage of your larvae. This can help you understand the expected range of variation and identify any potential genetic anomalies. If you are working with a species that has been extensively studied, there may be known genetic markers associated with specific larval traits. You can use this information to predict and interpret larval changes based on their genetic makeup. The point is: even in a controlled setup, genetic predispositions can still manifest and influence the larval development. Take notes when possible, and remember that understanding this

underlying genetic diversity is crucial for accurate interpretation.
Finally, when you collect enough data, the path forward involves compiling everything in a meaningful way. Raw data, photographs, temperature records, and food types are useless if not analyzed. Therefore, organize your notes and data into a format suitable for analysis. This could involve creating graphs, charts, or statistical models to identify trends and correlations. Look for patterns in the data that link larval changes to specific environmental factors or genetic traits. This might involve calculating growth rates, analyzing color distributions, or performing statistical tests to compare different treatment groups. It's time for you to use what you've observed and turn that into information. Remember, interpreting larval changes is not a one-time event but an ongoing process. Regularly review your data, refine your hypotheses, and adjust your experimental parameters as needed. Be prepared to adapt your approach based on new information or unexpected findings. The more you observe and analyze, the better you will become at understanding the complex and fascinating world of larval development. This is the heart of scientific discovery: a constant cycle of observation, analysis, and refinement.

6.3 Drawing Meaningful Conclusions

Drawing Meaningful Conclusions
The numbers are in, the larvae have morphed (or not),

and now stares the critical task: deciphering the story woven within the data. This stage often feels like entering a labyrinth, particularly after the meticulous detail demanded by controlled experiments. Don't be intimidated; the journey is about translation, transforming raw observations into understandable insights. To that end, a measured approach, fortified by skepticism and a willingness to challenge assumptions, is paramount. Let's move beyond simply crunching numbers; it's about extracting actionable knowledge.

Initial findings often raise more questions than answers, and the first step lies in grounding our interpretation of the data in the study's initial objectives. Return to your hypothesis, that initial proposition that drove the experiment. Does your data support, refute, or partially support it? Now is the time to honestly assess the findings without forcing them to fit a preconceived narrative. The beauty of scientific inquiry lies in its capacity to challenge established beliefs, and a 'failed' hypothesis can be just as valuable, if not more so, than a confirmed one. A simple, but often overlooked step, is to re-examine the raw data. Did any outliers skew the results? Were there any inconsistencies in the data collection process? Addressing these questions at this stage can save time and frustration later. Also, be mindful of the limitations of your experiment. Controlled environments, while offering precision, can sometimes oversimplify natural processes. The findings may not translate perfectly to real-world conditions, and acknowledging this limitation is crucial for responsible reporting.

Statistical significance is paramount, but it should not be the sole focus. A statistically significant result indicates that the observed effect is unlikely to have occurred by chance; however, it doesn't necessarily imply practical significance. For example, a minor change in larval development that is statistically significant might be of little biological importance. Instead, assess the magnitude of the effect size, which quantifies the practical importance of your findings. Common measures of effect size include Cohen's d or eta-squared. These metrics provide a standardized way to compare the magnitude of effects across different studies, allowing researchers to assess the real-world relevance of their findings. Moreover, confidence intervals offer a range of plausible values for the true population parameter. A narrow confidence interval suggests a precise estimate, whereas a wide interval implies greater uncertainty. By considering both statistical significance and effect size, researchers can gain a more comprehensive understanding of their results.

Comparative analysis is essential. How do your findings align with previous research in the field? Do they corroborate existing knowledge, challenge it, or offer a novel perspective? A thorough literature review is vital for contextualizing your results and identifying potential explanations for any discrepancies. Consider the methodology used in previous studies. Were the experimental conditions similar? Were the same species used? Differences in methodology can account for variations in results. Furthermore, consider the broader implications of your findings. How might they contribute to our understanding of egg-laying behaviour, larval development, or evolutionary processes? What are the potential applications of this knowledge in areas such as pest management, conservation, or biotechnology? By considering these broader implications, you can highlight the significance of your research and its potential impact on society.

Correlation does not equal causation, and this is a common pitfall in data interpretation. While correlation can provide valuable insights into the relationships between variables, it cannot establish a cause-and-effect relationship. For example, you might observe a correlation between temperature and egg-laying rate, but this doesn't necessarily mean that temperature is the sole cause of the change in egg-laying rate. Other factors, such as humidity, light, or food availability, could also be playing a role. Establishing causation requires carefully designed experiments that manipulate the independent variable (e.g., temperature) while controlling for all other potential confounding factors. Moreover, a thorough understanding of the underlying biological mechanisms is essential for supporting a causal claim. If there are multiple potential explanations for your findings, consider conducting further experiments to rule out alternative explanations and strengthen the evidence for your hypothesis.

Visual representations can illuminate complex data patterns. Charts, graphs, and diagrams are potent tools for showcasing your findings in an accessible and engaging manner. Choose the appropriate type of visualization based on the nature of your data and the message you aim to convey. For example, a scatter plot might be suitable for illustrating the relationship between two continuous variables, while a bar chart might be more appropriate for comparing the means of different groups.

Ensure your visuals are clear, concise, and easy to understand. Label axes appropriately, use informative captions, and avoid cluttering the image with unnecessary details. Visualizations should be used to complement, not replace, the statistical analysis. Highlight key trends and patterns in the data, but avoid over-interpreting the results. Visualizations are a powerful tool for communication, helping to convey the essence of your findings to a broader audience.

Transparency is critical, and the presentation of your findings should be as transparent and objective as possible. Clearly state your methodology, including any limitations or assumptions. Provide detailed descriptions of your data analysis techniques and any statistical tests used. Report both positive and negative findings, even if they don't support your initial hypothesis. Avoid cherry-picking data or selectively reporting results that align with your expectations. Be open to alternative interpretations of the data and acknowledge any uncertainties or inconsistencies. Transparency builds trust in your research and allows other scientists to evaluate the validity of your findings. Moreover, it promotes reproducibility, which is a cornerstone of the scientific method.

Replication is vital to confirm findings. Ultimately, the strength of any conclusion relies on its ability to be replicated by other researchers. If your findings cannot be replicated, it raises questions about the validity of your research and the reliability of your results. Encourage other scientists to repeat your experiments and verify your findings. Provide detailed protocols and datasets to facilitate replication efforts. Consider collaborating with other research groups to conduct multi-site studies, which can increase the generalizability of your findings. Replication is a crucial step in the scientific process, helping to ensure the robustness and reliability of scientific knowledge. Be prepared to refine your conclusions based on the results of replication studies. The path to meaning isn't always linear, and it is a process of continual refinement and adjustment. Sometimes, the initial conclusions will be altered as new evidence emerges. This is a natural and healthy part of the scientific process. Be open to changing your mind based on new information. Embrace the complexity of the natural world and avoid oversimplifying complex phenomena. Remember that scientific knowledge is always evolving, and the pursuit of understanding is a lifelong journey. By embracing this spirit of inquiry, researchers can contribute to the advancement of knowledge and improve the lives of others. The process of drawing meaningful conclusions is not merely a culmination of an experiment but rather a stepping stone towards further exploration and discovery. With each study, we deepen our understanding of egg-laying and larval transformations, paving the way for future advances in zoology, developmental biology, and beyond.

7 Chapter 7: Advanced Techniques & Future Directions

Now, the horizon expands, inviting us to contemplate the cutting edge of research and the exciting possibilities that lie ahead. It's time to move beyond the fundamental techniques and delve into methods that push the boundaries of our understanding of egg laying and larval development, methodologies that can be adapted from the world of genetics, imaging and more.

Genetic manipulations provide unprecedented power to dissect the molecular mechanisms governing these complex processes. Techniques like CRISPR-Cas9 gene editing have revolutionized our ability to target specific genes and observe the consequences on egg production, larval growth, and developmental pathways. For instance, we can knock out genes suspected of playing a role in yolk protein production and assess the impact on egg size, viability, and subsequent larval development. By examining the phenotypes that arise from these targeted mutations, we gain direct insights into the function of individual genes in these critical life stages. Gene knockdown using RNA interference (RNAi) is another powerful tool, allowing for transient reduction in gene expression to study the effects on egg-laying behavior or larval morphology. Consider the application of these techniques to investigate genes involved in hormone signaling pathways that regulate vitellogenesis (yolk formation); disrupting these pathways can reveal the intricate interplay between endocrine signals and reproductive success. Furthermore, introducing specific genes into organisms, referred to as transgenesis, enables researchers to study the effects of overexpressing particular proteins or introducing novel genetic elements, thus giving them a handle to manipulate genetic information, to then study the phenotypic effects, and understand how these phenotypic differences affect evolution and adaptation. All of these approaches should be done with the proper ethical approach.

The advancements in imaging technologies have opened up new vistas for visualizing and quantifying the intricate details of egg laying and larval transformations. Confocal microscopy allows us to create three-dimensional reconstructions of developing eggs, revealing the precise organization of cellular components and the dynamics of their rearrangement during embryogenesis. Time-lapse microscopy provides invaluable insights into the temporal sequence of events, capturing the dynamic changes that occur during egg activation, cell division, and tissue morphogenesis. Imagine tracking the movement of specific proteins within a developing embryo using fluorescently tagged antibodies, visualizing their localization and interactions in real-time. For example, light-sheet microscopy minimizes phototoxicity, enabling long-term imaging of delicate larval structures without causing significant damage. Magnetic resonance imaging (MRI) can be adapted to visualize internal structures of eggs and larvae non-invasively, allowing for longitudinal studies of growth and development. Moreover, powerful software tools enable quantitative analysis of image data, allowing us to measure cell sizes, track cell lineages, and quantify protein expression levels with unprecedented precision. The convergence of advanced imaging and computational analysis is transforming our ability to understand the complexities of development at the cellular and molecular level, unlocking insights into previously hidden processes, and allowing us to explore the secrets of evolution and adaptation of these organisms.

Looking ahead, the future of research in egg laying and larval transformations is brimming with exciting possibilities. One promising avenue is the integration of "omics" technologies – genomics, transcriptomics, proteomics, and metabolomics – to obtain a comprehensive understanding of the molecular events that underpin these processes. By analyzing the complete set of genes, transcripts, proteins, and metabolites present in eggs and larvae at different developmental stages, we can generate a holistic view of the molecular landscape and identify key regulatory networks. For example, single-cell RNA sequencing can reveal the gene expression profiles of individual cells within a developing embryo, providing insights into cellular differentiation and cell-cell communication. Another promising direction is the application of computational modeling to simulate developmental processes. By creating mathematical models that capture the interactions between genes, proteins, and environmental factors, we can predict the outcomes of experimental manipulations and gain a deeper understanding of the underlying mechanisms. The insights gained from these models can then be used to design new experiments and refine our understanding of development. Furthermore, research into the microbiome of egg-laying organisms and their larvae could reveal crucial interactions influencing development, immune function, and overall health. This exploration could offer novel insights into how microbial communities affect the host's physiology and evolutionary adaptations.

Moreover, the field stands to gain from interdisciplinary collaborations, integrating expertise from areas such as materials science, engineering, and nanotechnology. For example, new biomaterials could be developed to create artificial eggshells that mimic the properties of natural eggshells, allowing for precise control over the microenvironment of developing embryos. Nanotechnology could be used to deliver specific molecules or genes into eggs or larvae, providing a powerful tool for manipulating development. The confluence of these disciplines will propel the field forward, catalyzing new discoveries and fostering a deeper appreciation of the elegance and complexity of egg laying and larval transformations. By pushing the boundaries of our knowledge and embracing innovative approaches, we can unlock the secrets of these remarkable life stages and gain a more profound understanding of the intricate tapestry of life, its evolution and its adaptation. The integration of advanced statistical methods and machine learning will also prove invaluable. These tools can handle large, complex datasets generated by omics technologies and imaging techniques, enabling the discovery of patterns and correlations that might otherwise be missed. The application of machine learning algorithms can also aid in predicting developmental outcomes based on various parameters, thereby optimizing experimental design and improving the efficiency of research. The horizon is vast, and the potential for discovery is limitless.

7.1 Genetic Manipulations

Genetic manipulations, while presenting a realm of possibilities for understanding and influencing egg-laying and larval transformations, require careful consideration and expertise. For the beginner, the prospect can seem daunting, yet with a systematic approach and a focus on foundational concepts, the journey can be both rewarding and insightful. It's crucial to understand that these techniques are not about haphazardly altering genetic material but rather about precisely targeting specific genes or pathways to observe their effects on the developmental processes.

Before venturing into the complexities of gene editing, a firm grasp of basic genetics is essential. Understanding DNA structure, gene expression, and the mechanisms of inheritance forms the bedrock upon which all genetic manipulations are built. Start by familiarizing yourself with the model organisms you intend to study. What genes are known to influence egg production or larval development in that species? What are their functions? Publicly available databases like NCBI and model organism-specific repositories are invaluable resources. Delve into research papers and reviews to gain a comprehensive understanding of the genetic landscape. This initial phase should feel like detective work, piecing together the clues that will inform your experimental design. Moreover, recognize that ethical considerations are paramount. Always prioritize animal welfare, ensuring that any genetic manipulation is conducted responsibly and with the utmost care for the organisms involved. Obtain necessary approvals from institutional animal care and use committees (IACUCs) before initiating any experiments.

One of the most accessible entry points into genetic manipulation is through RNA interference (RNAi). This technique allows you to "knock down" the expression of a specific gene by introducing small interfering RNAs (siRNAs) into the organism. These siRNAs bind to the messenger RNA (mRNA) of the target gene, leading to its degradation and preventing the protein from being produced. For beginners, RNAi offers several advantages: it's relatively straightforward to implement, doesn't permanently alter the genome, and can be used to study the effects of gene knockdown on egg-laying or larval development. However, be aware of potential off-target effects. siRNAs can sometimes bind to unintended targets, leading to spurious results. Carefully design your siRNAs to minimize off-target binding and always include appropriate controls in your experiments to account for any non-specific effects. Moreover, remember that RNAi typically results in a reduction, not a complete elimination, of gene expression.

Another powerful tool in the genetic manipulation toolbox is CRISPR-Cas9. This technology allows for precise editing of the genome, enabling you to knock out genes, introduce specific mutations, or even insert new DNA sequences. While CRISPR-Cas9 is more complex than RNAi, it offers unparalleled precision and versatility. The CRISPR-Cas9 system consists of two main components: the Cas9 protein, which acts as a molecular scissor, and a guide RNA (gRNA), which directs the Cas9 protein to the specific target sequence in the genome. To use CRISPR-Cas9, you first need to design a gRNA that is complementary to the target sequence. The gRNA then guides the Cas9 protein to the target site, where it makes a double-stranded break in the DNA. The cell's natural repair mechanisms then kick in, either disrupting the gene or incorporating a new DNA sequence if provided. Before embarking on a CRISPR-Cas9 experiment, thoroughly research the potential off-target effects. The gRNA can sometimes bind to sequences that are similar but not identical to the target sequence, leading to unintended mutations in other parts of the genome. Use bioinformatic tools to identify potential off-target sites and design your gRNA to minimize off-target binding. Also, carefully consider the delivery method of the CRISPR-Cas9 components. Microinjection is a common technique for introducing the Cas9 protein and gRNA into eggs or early-stage embryos.

When designing any genetic manipulation experiment, meticulous planning is crucial. Clearly define your research question, identify the target gene or pathway, and select the appropriate technique (RNAi, CRISPR-Cas9, etc.). Include appropriate controls in your experiments to account for any non-specific effects. For example, a negative control should receive a non-targeting siRNA or gRNA that does not match any sequence in the organism's genome. Also, consider the potential for compensatory mechanisms. If you knock out a gene, the organism may compensate by upregulating other genes in the same pathway. Include multiple genes within a common pathway to prevent the compensatory upregulation. Replicates are your friend. Ensure you have sufficient biological replicates for statistical analysis. The more replicates, the more credible the data.

Following a successful genetic manipulation, the next step is to assess the effects on egg-laying or larval development. This may involve monitoring egg production, measuring larval growth rates, examining larval morphology, or analyzing gene expression patterns. Use a combination of techniques to obtain a comprehensive understanding of the phenotype. When analyzing the data, carefully consider the limitations of the technique. RNAi may not completely knock down gene expression, and CRISPR-Cas9 can sometimes lead to off-target effects. Always interpret the results in the context of these limitations. Don't jump to conclusions; remain critical about your analysis.

Genetic manipulations are powerful tools, but they are not without their challenges. Off-target effects, compensatory mechanisms, and technical difficulties can all complicate the interpretation of results. Therefore, patience and persistence are key. Don't be discouraged if your initial experiments don't go as planned. Learn from your mistakes and refine your approach. The field of genetic manipulation is constantly evolving, with new technologies and techniques emerging all the time. Stay up-to-date on the latest advances and be open to exploring new approaches. Finally, communicate your findings with the scientific community. Share your successes, as well as your failures, to help advance the field of egg-laying and larval transformation research. By embracing a systematic approach, a firm grounding in basic genetics, and a commitment to ethical research practices, beginners can successfully navigate the exciting world of genetic manipulations and unlock new insights into the mysteries of development.

7.2 Imaging Technologies

Microscopy stands as the cornerstone of imaging technologies when studying egg-laying and larval transformations. Its ability to reveal the intricate details of cellular and subcellular structures is invaluable. For beginners, understanding the different types of microscopy and their applications is paramount. Light microscopy, the most basic form, uses visible light to illuminate and magnify the sample. It's ideal for observing general morphology and developmental stages of eggs and larvae. However, its resolution is limited, which means it cannot resolve structures smaller than approximately 200 nanometers. To overcome this, advanced techniques like phase contrast and differential interference contrast (DIC) microscopy can be employed. These techniques enhance the contrast of transparent specimens without staining, allowing for detailed observation of internal structures within eggs and larvae.

Beyond basic light microscopy, fluorescence microscopy offers a powerful way to visualize specific molecules and structures within the sample. This technique involves labeling cellular components with fluorescent dyes or proteins, which then emit light when excited by a specific wavelength. Fluorescence microscopy can be used to track the expression of specific genes during development, observe the localization of proteins within cells, and visualize the dynamics of cellular processes. For example, one could use fluorescently labeled antibodies to track the expression of a specific protein involved in shell formation in eggs, or to observe the migration of cells during larval development. Confocal microscopy, a specialized form of fluorescence microscopy, further enhances image quality by eliminating out-of-focus light. This results in sharper, clearer images, especially when examining thick samples like whole eggs or larvae. Confocal microscopy allows for optical sectioning, which means that a series of images can be taken at different depths within the sample, allowing for three-dimensional reconstruction of the structure.

Electron microscopy takes visualization to an entirely different level. By using beams of electrons instead of light, electron microscopy achieves significantly higher resolution, allowing for the observation of structures at the nanometer scale. There are two main types of electron microscopy: transmission electron microscopy (TEM) and scanning electron microscopy (SEM). TEM involves transmitting a beam of electrons through a thin sample, creating an image based on the electron density of the material. TEM is ideal for examining the ultrastructure of cells and tissues, such as the arrangement of organelles within cells, or the structure of the eggshell. SEM, on the other hand, scans the surface of the sample with a focused beam of electrons, creating an image based on the electrons that are reflected or emitted from the surface. SEM is excellent for visualizing the external morphology of eggs and larvae, providing detailed information about their surface features and texture. While electron microscopy offers unparalleled resolution, it requires extensive sample preparation, including fixation, dehydration, and coating with a conductive material. It is also limited to examining dead samples, as the high vacuum environment required for electron microscopy is not compatible with living organisms.

Time-lapse imaging is not a specific type of microscopy but rather a technique that can be applied to various microscopy methods. It involves capturing a series of images over time, allowing for the observation of dynamic processes. Time-lapse imaging is particularly useful for studying egg-laying behavior, larval development, and cellular movements. For example, one could use time-lapse microscopy to observe the process of egg deposition, recording the precise timing and sequence of events. Or, time-lapse imaging could be used to track the migration of cells during gastrulation, providing insights into the mechanisms that drive development. When performing time-lapse imaging, it is important to consider factors such as the time interval between images, the exposure time, and the potential for phototoxicity. Phototoxicity refers to the damage caused to the sample by prolonged exposure to light, which can be a concern when using fluorescence microscopy. To minimize phototoxicity, it is important to use the lowest possible light intensity and to minimize the exposure time.

Optical projection tomography (OPT) is an emerging imaging technology that offers a unique way to visualize three-dimensional structures in whole organisms. OPT works by projecting light through the sample from multiple angles, and then reconstructing a three-dimensional image from the resulting projections. OPT is particularly well-suited for imaging small, transparent organisms, such as insect eggs and larvae. It allows for the visualization of internal structures without the need for sectioning, which can be time-consuming and introduce artifacts. OPT can be used to study the development of organs and tissues within the developing embryo, or to visualize the distribution of cells within the larval body. While OPT is a powerful technique, it requires specialized equipment and expertise.

Furthermore, choosing the right imaging technology also depends on the specific research question. If the goal is to observe the overall morphology of eggs and larvae, basic light microscopy or SEM may suffice. If the goal is to visualize specific molecules or structures within cells, fluorescence microscopy or TEM may be necessary. And if the goal is to study dynamic processes over time, time-lapse imaging should be considered. Regardless of the chosen technology, careful sample preparation is crucial for obtaining high-quality images. This includes fixation, staining, and mounting of the sample. The specific protocols for sample preparation will vary depending on the imaging technology and the organism being studied. As a final note, remember to consider the ethical implications of imaging technologies, particularly when working with living organisms. Minimize stress and harm to the animals, and adhere to all relevant ethical guidelines and regulations.

7.3 Future Research Avenues

The meticulous study of egg-laying and larval development in controlled environments offers fertile ground for future exploration. While current research provides a solid foundation, numerous avenues remain largely unexplored, promising groundbreaking discoveries across zoology, developmental biology, genetics, ethology, and evolutionary biology. One promising direction involves delving deeper into the genetic control of egg-laying behavior and larval metamorphosis, leveraging advanced genomic tools to identify and characterize the genes involved. The application of CRISPR-Cas9 technology, for instance, could enable precise gene editing to investigate the functional roles of specific genes in these processes. Consider, for example, manipulating genes involved in hormone synthesis and signaling during larval development to observe the resulting effects on metamorphosis timing and morphology. This approach could provide insights into the molecular basis of developmental plasticity and how environmental factors interact with genetic predispositions to shape larval trajectories. Further research is needed to understand the epigenetic modifications that influence gene expression during these critical life stages. By examining DNA methylation patterns, histone modifications, and non-coding RNA expression, we can gain a deeper appreciation of how environmental cues can induce long-lasting changes in development and behavior. This area of study would also benefit immensely from transcriptomic analyses, allowing for the comprehensive profiling of gene expression changes during egg development and larval metamorphosis, leading to the identification of novel genes and pathways involved.

Moreover, the interplay between the microbiome and host development offers a fascinating area of investigation. Larvae, particularly those in aquatic environments, are constantly exposed to a diverse array of microorganisms. It is becoming increasingly clear that these microbes can profoundly influence larval growth, development, and even behavior. The study of the microbiome's impact on immune development and competence in larvae is particularly relevant, potentially revealing insights into novel strategies for disease control and prevention in aquaculture and wildlife conservation. Research could focus on identifying specific microbial species or consortia that promote or inhibit larval growth and metamorphosis, and on elucidating the mechanisms by which these microbes exert their effects. Perhaps certain bacterial metabolites act as signaling molecules, influencing hormone production or gene expression in the host. The manipulation of the larval microbiome through probiotic supplementation or fecal microbiota transplantation could be explored as a means to optimize larval development and resilience. This is not only significant for basic biological understanding, but also has clear implications for applied fields like aquaculture, where larval survival is a major bottleneck.

The study of behavioral aspects during egg laying and larval transformations is an exciting area of research. We should conduct extensive observational studies using sophisticated tracking systems. Automated video analysis and machine learning algorithms would enable researchers to precisely quantify movement patterns, social interactions, and foraging behavior during both the egg-laying phase and throughout larval development. The effects of environmental factors, such as temperature, light, and food availability, on these behavioral parameters could be systematically investigated. An example of this could be examining the selection behavior of larvae exposed to different food sources, identifying the sensory cues they use to locate suitable prey, and assessing the impact of these preferences on their growth and survival.

Another crucial avenue for future research lies in the development of more sophisticated imaging techniques to visualize the intricate processes occurring during egg-laying and larval development. High-resolution microscopy, including confocal and light-sheet microscopy, can provide detailed images of cellular and tissue organization, allowing researchers to track cell movements and differentiation during metamorphosis. In addition to structural imaging, functional imaging techniques, such as calcium imaging and voltage imaging, can be used to monitor neural activity and signaling pathways in real-time. These techniques could be applied to study the neural circuits that control egg-laying behavior or to investigate the mechanisms of sensory processing in larvae. Furthermore, the integration of imaging data with computational modeling could provide a powerful tool for simulating and predicting developmental processes.

The application of "omics" technologies, in particular, promises to accelerate our understanding of the molecular mechanisms underlying egg-laying and larval development. Proteomics, for example, can be used to identify changes in protein expression during metamorphosis, providing insights into the regulatory networks that control this process. Metabolomics can be used to profile the metabolic changes that occur during egg development and larval growth, revealing the pathways that are essential for energy production and biosynthesis. The combination of genomics, transcriptomics, proteomics, and metabolomics, often referred to as multi-omics, provides a holistic view of the molecular landscape during these critical life stages. The integration of these datasets with advanced bioinformatics tools can reveal complex interactions between genes, proteins, and metabolites, leading to a more comprehensive understanding of the underlying biology.

Finally, the translation of basic research findings into practical applications represents a significant opportunity. For example, a deeper understanding of the factors that influence egg-laying behavior and larval survival could lead to improved aquaculture practices, increasing the efficiency and sustainability of seafood production. The knowledge gained from studying larval metamorphosis could also inspire the development of novel biomedical technologies, such as tissue engineering and regenerative medicine. Larvae possess remarkable regenerative abilities, and studying the mechanisms that underlie these abilities could provide clues for developing therapies to repair damaged tissues and organs in humans. Furthermore, the insights gained from studying the interactions between larvae and their environment could inform conservation efforts, helping to protect vulnerable species from the impacts of climate change and pollution.

www.ingramcontent.com/pod-product-compliance
Lightning Source LLC
LaVergne TN
LVHW020234210926
848481LV00016B/2188
* 9 7 9 8 8 6 9 0 3 4 4 6 5 *